RENEWABLE ENERGY: RESEARCH, DEVELOPMENT AND POLICIES

BATS AND WIND ENERGY

LITERATURE SYNTHESIS, ANNOTATED BIBLIOGRAPHY AND ASSESSMENT METHODOLOGY ON POPULATION IMPACT

Renewable Energy: Research, Development and Policies

Additional books in this series can be found on Nova's website under the Series tab.

Additional e-books in this series can be found on Nova's website under the e-book tab.

RENEWABLE ENERGY: RESEARCH, DEVELOPMENT AND POLICIES

BATS AND WIND ENERGY

LITERATURE SYNTHESIS, ANNOTATED BIBLIOGRAPHY AND ASSESSMENT METHODOLOGY ON POPULATION IMPACT

SEAN WILLIS
EDITOR

nova publishers
New York

NOTICE TO THE READER

Library of Congress Cataloging-in-Publication Data

ISBN: 978-1-63483-616-6

Published by Nova Science Publishers, Inc. † New York

CONTENTS

Preface		**vii**
Chapter 1	Bats and Wind Energy: A Literature Synthesis and Annotated Bibliography *Laura E. Ellison*	**1**
Chapter 2	Preliminary Methodology to Assess the National and Regional Impact of U.S. Wind Energy Development on Birds and Bats *J. E. Diffendorfer, J. A. Beston, M. D. Merrill, J. C. Stanton, M. D. Corum, S. R. Loss, W. E. Thogmartin, D. H. Johnson, R. A. Erickson and K. W. Heist*	**91**
Index		**171**

PREFACE

Since early on in the development of wind-energy production, concerns have arisen about the potential impacts of turbines to wildlife; these concerns have especially focused on the mortality of birds. Structural changes and improved turbine design have been instrumental in reducing mortality in birds. Despite the improvements to turbines that have resulted in reduced mortality of birds, there is clear evidence that bat mortality at wind turbines is of far greater conservation concern. Larger and taller turbines actually seem to be causing increased fatalities of bats. Numerous research opportunities exist that pertain to issues such as identifying the best and worst placement of sites for turbines; and mitigation strategies that would minimize impacts to wildlife (birds and bats). This book focuses on refereed journal publications and theses about bats and wind-energy development in North America.

In: Bats and Wind Energy
Editor: Sean Willis
ISBN: 978-1-63483-616-6

Chapter 1

BATS AND WIND ENERGY: A LITERATURE SYNTHESIS AND ANNOTATED BIBLIOGRAPHY*

Laura E. Ellison

EXECUTIVE SUMMARY

Turbines have been used to harness energy from wind for hundreds of years (Gipe, 2004). However, with growing concerns about climate change, wind energy has only recently entered the mainstream of global electricity production. In 2010, the global installed capacity for wind energy reached 196,630 Megawatt (MW), which represents approximately 2.5 percent of the total global energy consumption (World Wind Energy Report, 2010). In the United States, the total utility-scale wind power capacity through the 3rd quarter of 2011 totaled 43,461 MW and this represents more than 20 percent of the world's installed wind power (American Wind Energy Association, 2012). In 2011, the electricity produced from wind energy in the United States amounted to 120 Terawatt-hours (thousand MW) or 2.9 percent of total global electricity demands (U.S. Energy Information Administration, 2012). Canada is the ninth largest producer of wind energy in the world with current installed

* This is an edited, reformatted and augmented version of U.S. Geological Survey Open-File Report 2012–1110, issued in 2012.

capacity at 4,862 MW, representing about 2.1 percent of Canada's total electricity demand (Canadian Wind Energy Association, 2012).

Since early on in the development of wind-energy production, concerns have arisen about the potential impacts of turbines to wildlife; these concerns have especially focused on the mortality of birds. Early styles of turbines appeared to pose a greater risk to birds in terms of collision mortality than more modern turbines do (Erickson and others, 2002; Young and others, 2003). Early turbines were smaller, had a higher blade-rotation rate, and had a lower energy output. This resulted in more turbines being needed for significant electricity production, thereby increasing the chances of birds encountering turbines (Curry and Kerlinger, 2000; Erickson and others, 2002; Howell, 1995). The lattice towers of smaller turbines also provided birds with perching opportunities, which was thought to further increase mortality (Kerlinger, 2002; Orloff and Flannery, 1992; Osborn and others, 1998). Structural changes and improved turbine design have been instrumental in reducing mortality in birds (Johnson and others, 2002; Smallwood and Karas, 2009). For example, during a study at the Altamont Pass Wind Resource Area in California, it was found that when comparing the concurrently operating old-generation, smaller turbines during 2005–2007, adjusted fatality rates in the newer, larger, and taller turbines were 66 percent lower for all birds combined.

Despite the improvements to turbines that have resulted in reduced mortality of birds, there is clear evidence that bat mortality at wind turbines is of far greater conservation concern. Larger and taller turbines actually seem to be causing increased fatalities of bats (Barclay and others, 2007). Bats of certain species are dying by the thousands at turbines across North America, and the species consistently affected tend to be those that rely on trees as roosts and most migrate long distances (Cryan and Barclay, 2009). Bat mortality at wind-energy facilities was first documented in Australia, where 22 white-striped mastiff-bats (*Tadarida australis*) were found at the base of turbines over 4-year (yr) period (Hall and Richards, 1972). In 1999, 45 dead bats were found at a wind energy facility in Carbon County, Wyoming; 10 dead bats were found at a wind energy facility in Umatilla County, Oregon; and 34 dead bats were found within a wind energy facility in Wisconsin (Keeley and others, 2001). Small numbers of dead bats have also been found at wind-energy facilities in California (Orloff and Flannery, 1992; Howell, 1997; Anderson and others, 2000; Thelander and Rugge, 2000). Turbine-related bat mortalities are now affecting nearly a quarter of all bat species occurring in the United States and Canada. Most documented bat mortality at wind-energy facilities has occurred in late summer and early fall and has involved tree bats, with hoary bats (*Lasiurus cinereus*) being the most prevalent among fatalities.

Populations of bats are difficult to monitor (O'Shea and others, 2003). Because of this, there is insufficient information on the population status of the 45 species of bats in the United States, especially for migratory foliage- and tree-roosting bats (O'Shea and others, 2003). With this lack of understanding of total population sizes, demographics, and impacts of fatalities from wind turbines on the viability of affected bat populations, it is currently not possible to determine the influence of any single source of mortality or of any effects of mitigation strategies on these bat populations. In addition to the direct effects of wind-energy development on bat mortality, indirect effects may occur as well. Bats have low reproductive rates and generally give birth to a single individual once a year. This results in bat populations growing slowly and an inability to quickly rebound after rapid declines in population size. Bat populations therefore rely on high adult survival rates to compensate for low reproductive rates and prevent declines. Therefore, substantial cumulative impacts of wind-energy development on certain bat species, especially tree-roosting bats, are expected, and these populations would be slow to recover from any population declines (Barclay and Harder, 2003).

Numerous research opportunities exist that pertain to issues such as: (1) identifying the best and worst placement of sites for turbines and (2) mitigation strategies that would minimize impacts to wildlife (birds and bats). Unfortunately, to date, very little research of this type has appeared in the peer-reviewed scientific literature; much of the information exists in the form of unpublished reports and other forms of gray literature. This literature synthesis and annotated bibliography focuses on refereed journal publications and theses about bats and wind-energy development in North America (United States and Canada). Thirty-six publications and eight theses were found, and their key findings were summarized. These publications date from 1996 through 2011, with the bulk of publications appearing from 2007 to present, reflecting the relatively recent conservation concerns about bats and wind energy.

The idea for this Open-File Report formed while organizing a joint U.S. Fish and Wildlife Service/U.S. Geological Survey "Bats and Wind Energy Workshop," on January 25–26, 2012. The purposes of the workshop were to develop a list of research priorities to support decision making concerning bats with respect to siting and operations of wind-energy facilities across the United States. This document was intended to provide background information for the workshop participants on what has been published on bats and wind-energy issues in North America (United States and Canada).

FORMAT OF REPORT

This report is divided into three sections: (1) a literature synthesis; (2) an annotated bibliography; and, (3) additional references. The literature synthesis and annotated bibliography focus on North America and on refereed journal publications. Additional references include a selection of citations on bat ecology, international research on bats and wind energy, and unpublished reports.

LITERATURE REVIEW METHODS

A detailed literature review was conducted using Internet resources and databases. The keywords chosen for these searches included "bats," "Chiroptera," "wind," and "wind energy."

Keywords were used alone or in combination with the other terms. Databases and search engines used included Google (*http://www.google.com*), Google Scholar (*http://scholar.google.com*), SciVerse Scopus (*http://www. scopus. com*), ISI Web of Knowledge (*http://apps.isiknowledge. com*), and the USGS Library's Digital Desktop (*http://library.usgs.gov*). The literature-cited sections of publications obtained from keyword searches were also cross-referenced to identify additional citations or gray literature that were missed by the Internet search engines.

ACKNOWLEDGMENTS

The work on which this Open-File Report is based was conceived while organizing a joint U.S. Fish and Wildlife Service (FWS)/U.S. Geological Survey (USGS) "Bats and Wind Energy Workshop," on January 25–26, 2012. I thank the participants of the workshop for their suggestions and valuable input on the development of this document: From FWS, Paul Barrett (Southwest, R2), Gabriela Chavarria (Washington), Jeremy Coleman (Northeast, R5), Megan Cook (Washington), Craig Hanson (Mountain-Prairie,

R6), David Kampwerth (Pacific, R1), TJ Miller (Midwest, R3), Scott Pruitt (R3), and Jennifer Szymanski (R3). From USGS, Laurie Allen (USGS Ecosystems Mission Area), Dan Manier [Fort Collins Science Center (FORT)], Nina Burkhardt (FORT), Paul Cryan (FORT), Robb Diehl (Northern Rocky Mountain Science Center), Jay Diffendorfer (Rocky Mountain Geographic Science Center), Marcos Gorresen (Pacific Islands Ecosystem Research Center), Manuela Huso [Forest and Rangeland Ecosystem Science Center (FRESC)], Doug Johnson (Northern Prairie Wildlife Research Center), Sue Phillips (FRESC), and Wayne Thogmartin (Upper Midwest Environmental Sciences Center). Ed Arnett (Bat Conservation International), Mark Hayes (FORT), and Patty Stevens (FORT) provided helpful peer review comments. Jenny Shoemaker (USGS, FORT) assisted with formatting the final report. Funding for this Open-File Report was provided by Laurie K. Allen and William A. Lellis (USGS Ecosystems Mission Area).

I. Literature Synthesis

I found 36 refereed journal articles and eight theses during this literature search. The earliest article was published in 1996 (Osborn and others, 1996) and the number of publications per year increased until the present year. There was one publication in 1996, one in 2002, one in 2003, two in 2004, one in 2005, six in 2007, seven in 2008, four in 2009, six in 2010, nine in 2011, and finally, one in 2012. Five of the eight theses were annotated and included in this synthesis because they were intended for publication. Three of the theses were later published and annotated in this report. I identified five broad categories of topics: (1) Causes of mortality; (2) Fatalities and/or activity patterns; (3) Migration; (4) Mitigation and curtailment; and, (5) Review papers and syntheses. I summarize the literature for each of these topics in the following five subsections.

Causes of Mortality

Eight papers addressed the causes of bat mortality at wind-energy facilities. Kunz and others (2007a) and Cryan and Barclay (2009) provided overviews of the hypothesized causes of bat fatalities at turbines. These hypotheses fall into two categories: proximate and ultimate. Proximate causes explain the direct means by which bats die at turbines and include bats colliding with turbine towers, colliding with rotating blades, or barotrauma (that is, internal injuries suffered after being exposed to rapid pressure changes near the trailing edges and tips of the moving blades). Hypotheses of ultimate causes are numerous and include three general categories: random collisions, coincidental collisions, and collisions that occur because bats are attracted to turbines. Random collisions are those that occur due to chance alone or involve no assumptions of circumstance or attraction. Coincidental collisions involve bats being victims of unfortunate behavioral circumstances that put them at risk of colliding with turbines (for example, turbines are located along migratory pathways). Hypotheses of attraction propose that there is some attractor or combination of attractors drawing bats to wind turbines (Kunz and others, 2007).

The obvious prominent cause of bat deaths at wind turbines are direct collision (that is, blunt-force trauma) and barotrauma. Baerwald and others (2008) proposed that barotrauma was a significant cause of bat fatalities at wind turbines in southwestern Alberta, Canada. They found that 46 percent of bats killed at turbines had no discernible external injuries that would have been fatal, and 92 percent of bats necropsied had hemorrhaging in the thoracic and/or abdominal cavities. In a follow-up study at this same location, a small percentage (2.5 percent) of bats found while conducting fatality searches were alive and only 30 percent of those live bats had visible signs of skeletal damage or considerable soft tissue damage. However, Grodsky and others (2011) used veterinary diagnostic procedures to investigate bat fatalities in southeastern Wisconsin and found that the exact cause of death (that is, barotrauma or direct collision) could not be determined in most bats due to the variability of injuries and a lack of exclusively attributable lesions. They concluded that the cause of death for bats killed at turbines was not exclusively or predominantly barotrauma or direct collision but rather an indiscernible combination of both. Simply using a visual inspection of a bat carcass is not adequate for conclusively diagnosing fatal injuries, including broken bones. Rollins and others (2012) examined the causes of lung damage from salvaged bats and found that results could be confounded by common postmortem artifacts (for example, decomposition, temperature differences) as well as the competing hypothesized causes of traumatic injury (for example, blunt force trauma, barotrauma, laceration). Because these competing factors could not be confidently excluded during a pathological examination, pulmonary barotrauma should not be diagnosed from salvaged bat carcasses unless assessments are also made for ruptured ear drums. The authors conclude that their results point to a need for improvements in the current monitoring procedures at wind-energy facilities and in the experimental approaches used to determine whether barotrauma or other causes contribute to bat mortality.

Horn and others (2008) used thermal infrared (TIR) cameras to monitor the activity of bats at Mountaineer Wind Energy Center in West Virginia. They found clear evidence of bats avoiding turbine blades and of bats being directly struck by blades. One of the most important observations from this study was that bats were observed actively investigating both moving and motionless turbine blades; this suggests that they may be attracted to the turbines as potential roosting trees.

Kunz and others (2007) first proposed the hypothesis that tree bats are attracted to turbines. Cryan (2008) further elaborated on this hypothesis by proposing that tree bats collide with turbines while engaging in mating behaviors that center on the tallest trees in the landscape. Tree bats may use the highest trees in a landscape as rendezvous points during the mating season and this behavior places them at risk of collision with the blades of wind turbines. Jameson's (2010) thesis called this the "Reproductive Landmarks Hypothesis" and found evidence that tree bats were attracted to tall structures during the fall migration, the same time of year of most documented mortality at wind-energy facilities. Direct evidence of mating at communication towers and turbines was not observed during this study; however, high levels of bat activity were recorded at communication towers during migration.

Fatalities and/or Activity Patterns

Eleven papers and five theses investigated patterns of fatality and/or activity of bats at wind-energy facilities. Four papers and one thesis looked at fatalities alone, while the remaining studies examined both bat activity at and near turbines and bat mortalities. Patterns that emerged from these fatality and activity studies were the following: (1) fatality rates for bats at wind turbines and the species most affected differed by geographical region; (2) there was a tendency for more males to be killed than females (with a few exceptions); (3) most mortality occurred during fall migration; (4) the majority of fatalities were tree bats (hoary, silver-haired, and eastern red bats); (5) searcher detection and scavenger rates for bat carcasses are highly variable and depend on the region, vegetation, and methodology used; (6) there was no consistency in how mortality rates were presented; (7) bat-activity (acoustical) patterns around turbines and in nearby habitats varied by geographical region; and, (8) there was a tendency for bat fatalities to increase in low wind speeds.

Prior to Osborn and others (1996), most studies on the biological impacts of wind energy were designed to assess the impacts to birds. Bat collisions with turbines were infrequent and had not received attention by natural resource agencies or the scientific community. While conducting a study to assess impacts to birds from wind turbines at the Buffalo Ridge Wind Resource Area in southwestern Minnesota in 1994–1995, Osborn and others (1996) found 13 bat carcasses of 5 different species, 11 of which were tree bats (hoary, silver-haired, and red bats). A high percentage of these fatalities (85 percent) occurred during the summer. Johnson and others (2003) continued to document mortality of bats at this same location from 1996 to 1999 and found 163 fatalities: 89 percent of the fatalities were hoary and eastern red bats, and 97 percent occurred from July 15 to September 15. In 2001 and 2002, Johnson and others (2004) continued to conduct fatality searches as in the previous two studies, but included acoustic and mist-netting surveys to investigate bat activity and composition in relation to the wind facility. They found more bat activity at foraging and roosting areas than at turbines. Mortality of bats at wind turbines did not appear to involve foraging bats, but instead corresponded with post-breeding southward migration of hoary (*Lasiurus cinereus*), eastern red (*L. borealis*), and silver-haired bats (*Lasionycteris noctivagans*).

At about the same time as the work being done at Buffalo Ridge, Minnesota, Gruver (2002) conducted a study examining bat community structure and bat fatalities at the Foote Creek Rim Wind Power Facility in south-central Wyoming from 2000 to 2001. He found that although bats in the genus *Myotis* represented approximately 81 percent of all bats captured using mist nets, bat fatalities consisted of 22 hoary bats and 1 big brown bat over the course of this study. (Hoary bats and big brown bats are in different genera than *Myotis*.)

Previous to 2003 or 2004, mortalities were reported as numbers of carcasses found and were summarized by species. Around 2004, fatality surveys began to address the sampling and statistical issues of estimating mortality using carcass surveys. Studies began to incorporate searcher efficiency and scavenger removal trials into their methodology. However, there was still a lot of variability in how mortality statistics were presented. For example, some studies lumped tree bats and reported an adjusted mortality rate for the year and for the entire location (Fiedler, 2004), some reported a rate of bats killed for the sampling period of the study [(for example, from

April to December 2003; Jain and others (2011)], and some reported the number of bats killed per turbine per day and/or per MW per day (Grodsky, 2010).

Bat mortality was investigated at Buffalo Mountain Windfarm in eastern Tennessee from 2000 to 2003 (Fiedler, 2004). At this location, eastern red bats made up the majority of bat fatalities (61.3 percent), followed by tricolored bats (22.4 percent), and hoary bats (10.1 percent). An adjusted mortality rate for this study and location was estimated to be 20.82 bats/turbine/year. On nights where fresh fatalities occurred at turbines, bat activity levels, determined by acoustic monitoring, were greater than on nights where no fatalities occurred. However, turbine collisions of red and hoary bats were disproportionately greater than the acoustical monitoring would have indicated if acoustic activity was directly proportional to fatality risk.

Jain and others (2011) examined bat mortality and activity at Top of Iowa Wind Resource Area in north-central Iowa during 2003–2004. They calculated an adjusted mortality rate for the entire wind-resource area for each year of the study. In 2003, mortality was estimated to be 396.05 plus or minus 71.71 (95 percent CI) bats (4.45 bats/turbine, 4.94 bats/MW) and in 2004, that estimate increased to 635.83 plus or minus 111.89 (95 percent CI) bats (7.14 bats/turbine, 7.94 bats/MW). There was no difference between bat activity at turbine sites and adjacent crop fields without turbines during both years.

Another Midwestern study took place in Wisconsin. Grodsky (2010) studied bird and bat mortality at the Forward Energy Center in southeastern Wisconsin. Daily fatality rates were calculated for migratory tree bats as 0.158 bats/turbine/day and 0.105 bats/MW/day. In addition to daily fatality rates, the author also corrected estimates for tree bat mortality to the number of bats per turbine and bats per MW over the entire study in order match the standards from other studies (Jain and others, 2007; Kerlinger and others, 2007; Gruver and others, 2009). These estimates were 47.865 bats/turbine/entire study and 31.875 bats/MW/entire study. By his estimation, approximately 4,454 total bats, 3,019 migratory bats, and 912 non-migratory bats were killed during the 2 springs and 2 falls of the study.

Piorkowski and O'Connell (2010) studied the spatial patterns of summer bat mortality from collisions with wind turbines at the Oklahoma Wind Energy Center in the southern Great Plains, Oklahoma. This study provided the first evidence of a steady rate of collision mortality of Brazilian free-tailed bats (*Tadarida brasiliensis*). Eighty-five percent of all bat fatalities were Brazilian free-tailed bats and rates of mortality adjusted for probability of detection and incorporating published removal estimates yielded an estimate of 74–92 individuals (0.73–0.90/MW) in 2004 and 67–85 (0.66– 0.83/MW) in 2005.

Until Miller's (2008) thesis, no studies of bat mortality at wind-energy facilities had been completed in Texas, despite Texas leading the United States in installed wind-energy capacity. Miller (2008) looked at patterns of avian and bat mortality at the Red Canyon Wind Energy Center in the southern portion of the Texas Panhandle from September 2006 to September 2007. As with Piorkowski and O'Connell's (2010) Oklahoma study, the Brazilian free-tailed bat represented the majority of all bat fatalities (94 percent), followed by hoary bats (4 percent), and eastern red bats (2 percent). The author classified turbines into three categories: low, medium, and high mortality. Turbines were assigned these three mortality categories based on elevation, aspect, and percent of the turbine site that was a part of the Caprock Escarpment edge, a distinct palisade-like escarpment. High mortality turbines were southwest-facing, less than 853 meters (m) in elevation with greater than 38 percent Caprock Escarpment edge within the turbine site boundary. Medium mortality turbines were south-facing, between 853 and 865 m in elevation, and with between 34 and 38 percent of the site boundary as Caprock edge. Low mortality turbines were southeast-facing, greater than 865 m in elevation, and less than 34 percent of the site boundary as Caprock edge. Low mortality turbines were estimated to kill 15.29 bats/MW, medium mortality turbines were estimated to kill 45.00 bats/MW, and high mortality turbines were estimated to kill 106.67 bats/MW. Total mortality was estimated for the entire wind energy development to be 36.87 bats/MW.

At the Altamont Pass Wind Resource Area (APWRA) in California, Smallwood and Karas (2008) examined avian and bat mortality rates at older-generation turbines compared to repowered (larger and taller) turbines over several years (1998–2003 and 2005–2007). Adjusted fatality rates for bats increased from zero at the old vertical-axis turbines to 16.4/year at the new, repowered turbines. This was nearly 800 percent greater bat mortality at the new, larger turbines when compared to the older-generation turbines. However, the newer turbines were found to reduce bird fatality rates a great deal. The authors extrapolated the average adjusted bat fatality rate from the repowered turbines to a completely repowered APWRA and estimated about 454 bat fatalities/year might result.

Weller and Baldwin (2011) combined the results of continuous echolocation and meteorological monitoring at multiple stations to model the conditions that explain the presence of migratory bats at a wind energy facility in southern California. Despite low overall bat-activity levels, the authors were able to build models that successfully predicted conditions when migratory bats were present. Using acoustic detectors and occupancy modeling, they found that occupancy of bats was associated with lower nightly wind speeds and higher nightly temperatures, mirroring results from other wind-energy facilities in other regions and habitats. The results also highlight the importance of modeling conditions that explain bat activity on a seasonal basis. There were marked differences in activity levels seasonally; the relative importance of explanatory variables also differed seasonally. The authors conclude that using this model-based approach could improve the effectiveness and efficiency of mitigations for bat fatalities at wind-energy facilities. If properly conducted, echolocation monitoring and meteorological monitoring may provide a reliable index to expected bat fatality levels at wind-energy facilities.

Canadian bat mortality studies report relatively high fatalities at wind-energy facilities (Baerwald and Barclay, 2009; 2011; Jameson, 2010; Klug and others, 2011). Baerwald and Barclay (2009; 2011) examined the geographical variation in activity levels and fatality rates of bats at one proposed and eight existing wind-energy facilities across southern Alberta, Canada, from 2006 to 2007. Corrected fatality rates varied among the 9 sites sampled, from 31.41 plus or minus 3.04 bats/turbine/year to 1.26 plus or minus 0.18 bats/turbine/year, with higher rates at western wind-energy facilities and lower rates at eastern facilities. Fatality rates did not appear to be influenced by migratory bat activity at ground level or at 30 m. Jameson (2010) examined the factors influencing bat mortality at the St. Leon Wind Energy Project in south-central Manitoba, Canada, in 2008 and 2009. In 2008, adjusted mortality rates reported for this study were 11.5 plus or minus 4.9 bats/turbine (7.2 plus or minus 3.1 bats/MW) for silver-haired bats, 15.7 plus or minus 5.5 bats/turbine (9.9 plus or minus 3.4 bats/MW) for hoary bats, and 2.0 plus or minus 1.0 bats/turbine (1.2 plus or minus 0.6 bats/MW) for eastern red bats. In 2009, adjusted mortality estimates were 7.7 plus or minus 2.1 bats/turbine (4.6 plus or minus 1.3 bats/MW) for silver-haired bats, 5.8 plus or minus 2.1 bats/turbine (3.5 plus or minus 1.3 bats/MW) for hoary bats, and 1.1 plus or minus 0.8 bats/turbine (0.7 plus or minus 0.5 bats/MW) for eastern red bats.

Fatality Search and Estimation Techniques

Arnett (2006) conducted a baseline effort to assess the efficiency of dog-handler teams to recover bat fatalities at two wind-energy facilities in West Virginia and Pennsylvania. Overall dog-handler efficiency (percent of trial bats found) for all trials and bats combined, and using combined findings from both dogs, was 71 percent at Mountaineer (W. Va.) and 81 percent at Meyersdale (Pa.), compared to 42 percent and 14 percent for humans, respectively. Significant variability was found in searcher efficiency (both dog-handler and humans) based on vegetative cover, terrain, and amount of high-visibility habitat found at the two sites. Although the findings from this pilot study were promising, the author suggested that more research was needed to better understand patterns and to account for limitations and biases influencing the efficiency of dog-handler teams as a tool to improve fatality estimates for bats at wind-energy facilities.

Huso (2010) addressed the problem of estimating numbers of wildlife fatalities from observed carcasses. She proposed a fatality estimator that accounts for searcher efficiency, scavenger removal, search interval, density-weighted area searched, and visibility within search transects. The proposed estimator's bias and precision were compared to two other estimators in common use during bat-carcass surveys. Bias is the difference between a population mean of the estimator and an accepted reference or true value (Bainbridge, 1985). Precision is the measure of the statistical variance of an estimation procedure (West, 1999). None of the three estimators compared were unbiased under all conditions. However, bias in the proposed estimator never exceeded plus or minus 27 percent, whereas bias in the other two estimators was always negative and exceeded that of the proposed estimator in 98 percent and 93 percent of the simulated conditions, respectively. The proposed estimator was relatively robust to variation in sources and magnitudes of imperfect detectability, but was sensitive to distributional assumptions regarding carcass-removal rates and searcher efficiency. The proposed estimator offers significant improvement over two current estimators and provides relatively unbiased estimates of fatality that can be applied under a variety of conditions and survey protocols.

Migration

Four publications examined migration patterns of bats in relation to wind-energy facilities. A fifth paper provided a short review of how certain aspects of bird migration could be useful for designing research on bat migration and wind-energy interactions (Larkin, 2006). Reynolds (2006) sampled spring migration at the Maple Ridge Wind Project in the Tug Hill Plateau ecoregion of western New York in 2005. He sampled migratory activity using acoustic monitoring at two sites and found that most of the migratory bat activity occurred at minimum wind speeds below 1.3 meters per second (m/s). He also found that days with high bat-migration activity had a mean maximum temperature of 23.9 plus or minus 4.4^{0}C and lower bat-migration activity at temperatures of 9.8 plus or minus 4.8^{0}C. Acoustic-monitoring results from this study strongly suggested that the Tug Hill Plateau region did not support a large bat population in terms of either species diversity or total bat abundance. Most of the migratory bat activity occurred at wind speeds below the cut-in speeds of a typical commercial wind turbine.

However, this study did not sample fall migration: the period in which most bat fatalities occur at wind-energy facilities, including the high fatality rates observed at the Maple Ridge Wind Farm where Reynolds monitored.

Cryan and Brown (2007) investigated migration of bats at the Southeast Farallon Island off the coast of California west of the Bay Area. They analyzed a 38-year (yr) data set of observations of hoary bats arriving and departing on this island. They also investigated whether moonlight and weather conditions influenced the occurrence of bats on the island. They found that hoary bats were observed more frequently in September than any other month. Bats were more likely to arrive and depart from the island on evenings when wind speeds were relatively low, during darker phases of the moon, and during overcast nights. These results are relevant to the problem of bat fatalities at wind turbines because (1) arrivals of migrating tree bats at particular sites may be predictable events, (2) there was conclusive evidence that tree bats migrate over offshore water where future construction of wind turbines has been proposed, (3) migratory hoary bats use vision to navigate during migration and are drawn toward visual stimuli, (4) there was a clear relationship between the timing of autumn migration by hoary bats and collisions with anthropogenic structures, and (5) there was a possible behavioral explanation for bat collisions with wind turbines.

Johnson and others (2011a) examined the seasonal bat activity on Assateague Island National Seashore off the coast of Maryland, in 2005 and 2006. They used acoustic monitoring stations at three locations on the island to sample bat activity. More than half the total bat passes were from eastern red bats (59.33 percent), followed by big brown bats (3.05 percent), hoary bats (0.10 percent), tricolored bats (0.10 percent), and silver-haired bats (0.07 percent). They found that bat activity generally increased in April, peaked in August, and declined gradually until December. Bats were less active during periods of low temperatures and high wind speeds, possibly above thresholds for wind turbine cut-in speed. The results from this study provided evidence that bats migrate along the Atlantic coastline and use barrier islands, possibly for navigation or as stopover sites, and this has implications for proposed offshore wind-energy facilities.

Johnson and others (2011b) examined seasonal and geographic trends in acoustic detection of migrating tree-roosting bats at one existing and thirteen proposed wind-energy facilities in seven eastern states in the United States between April and November of 2007 and 2008. Acoustic surveys were placed into one of three geographic regions to allow for latitudinal comparisons: northeast, northern Allegheny Plateau, and mid-Atlantic. Nightly silver-haired, hoary, and eastern red bat calls/detector/night at two proposed sites in West Virginia were also compared to the number of mortalities at Mount Storm Wind Energy Facility, W. Va., using linear correlations. The authors found that nightly detection rates of silver-haired and hoary bats at the two sites in West Virginia were positively correlated with the number of dead silver-haired and hoary bats at the Mount Storm Wind Facility. However, no significant correlation was found between detection rate of eastern red bats and the mortality at Mount Storm. The nightly detection rates of silver-haired, hoary, and eastern red bats during fall migration peaked earlier in the year in the two northernmost geographic regions compared to the mid-Atlantic region. Conversely, nightly detection rates of the three species during the spring migration peaked earlier in the year in the mid-Atlantic region, compared to geographic regions farther north. In all regions, eastern red bats were more frequently detected later in the year than were hoary bats. This has implications for wind-energy facilities, in that the seasonal detection rates of hoary bats, silver-haired bats, and eastern red bats likely reflect their different migratory patterns, and these patterns may be useful in predicting the timing of mortality events at turbines.

Mitigation and Curtailment

Two publications examined mitigation and curtailment at wind-energy facilities to reduce the number of bat fatalities. Baerwald and others (2009) conducted a mitigation experiment at a wind energy facility in southwestern Alberta, Canada. Normal operating turbines at this facility had a cut-in speed of 4 m/s: the wind speed at which the turbine begins to generate electricity. The authors specifically tested the hypothesis that reducing the amount that turbine rotors turn in low wind speeds will reduce bat fatalities. They operationally altered randomly chosen turbines in two different ways. Fifteen turbines had their rotor start-up speed increased to 5.5 m/s, meaning that the turbines were idle and motionless during low wind speeds. Another six turbines were manipulated to change the pitch angle of their blades to reduce rotor speed in low wind, thereby lowering the generator speed required to start energy production, which caused turbines to be motionless similar to the other experimental treatment but with different implications for turbine operations. Both sets of experimental turbines killed fewer bats than did the control (normal operating) turbines, and there was no difference between the two experimental treatments. The corrected fatality rates for each species of migratory bat were reduced by between 50 percent and 70 percent at the experimental turbines. The effect of changes in operation was different between the two experimental turbines, with the increased rotor start-up speed (5.5 m/s) being more costly than changing the pitch angle of the blades.

Arnett and others (2011) also tested the effectiveness of raising the wind-turbine cut-in speed to reduce bat fatality. Wind-turbine cut-in speed was defined to be the lowest wind speed at which turbines generate power to the utility system. Raising the wind-turbine cut-in speed renders turbines non-operational until the higher cut-in speed is reached; then turbines begin to spin and produce power. This study took place at the Casselman Wind Project in Somerset County, Pa., during 2008 and 2009. Three turbine treatments were defined and assigned (in a randomized block design) to 12 turbines. Treatments were that turbines became operational and began to spin with (1) a cut-in speed of 3.5 meters per second ($m\ s^{-1}$; normal operations), (2) cut-in speed of 5.0 $m\ s^{-1}$, and (3) cut-in speed of 6.5 $m\ s^{-1}$. Treatment-related mortality was measured as the sum of all individual carcasses of bats estimated to have been killed during the previous night observed along transects near the turbines. The results of this study found no difference between the number of fatalities for the two raised cut-in speed treatments for both years. Observed bat mortality at fully operational turbines was, on average, 5.4 (2008) and 3.6 (2009) times greater than mortality associated with curtailed (that is, non-operating) turbines. The authors conclude that relatively small changes to wind-turbine operation resulted in nightly reductions in bat mortality, ranging from 44 percent to 93 percent, with marginal annual power loss (less than or equal to 1 percent of total annual output).

In both of the curtailment studies described above, all species of bats were combined, and species-specific effects of curtailment are unknown. Most fatalities at these curtailment sites were migratory tree bats.

Review Papers and Syntheses

Eleven papers found during this literature search were reviews or syntheses of bats and wind-energy issues. The first review paper published was in 2005 (Johnson, 2005). As of 2004, only 11 studies had been conducted in the United States to estimate fatality rates of bats at wind farms (Johnson, 2005). The numbers of fatality studies continued to increase in succeeding years. Five synthesis and technical guidance papers were published in 2007 (Arnett and others, 2007; Barclay and others, 2007; Kunz and others, 2007a; 2007b; Kuvlevsky and others, 2007). By 2006, Arnett and others (2008) were able to summarize the results from 21 studies of patterns of bat fatalities at 19 wind-energy facilities across North America. The patterns of bat fatalities that emerged from these various syntheses provided the scientific community the ability to construct numerous hypotheses about the causal reasons bats collide with turbines. These hypotheses and their resulting predictions were summarized by Cryan and Barclay (2009). Sovacool (2009) compared bird fatalities from wind energy, fossil fuel, and nuclear electricity in an attempt to contextualize the negative effects to bird populations. According to the author, fossil-fueled facilities are about 17 times more dangerous to birds on a per Gigawatt hour (GWh) basis than wind and nuclear power stations. Bats were included in the title of this paper; however, they were lumped with birds, and their unique life-history strategies were not considered in the paper. Willis and others (2010) provided a response to Sovacool (2009) that examined the various problems with lumping birds and bats together when comparing mortality resulting from the three methods of electricity generation. The most recent review publication was Cryan (2011), and his review summarizes bat mortality caused by wind turbines and the potentially significant effects of this mortality on certain bat populations.

The following general patterns and findings common to all of the review papers are listed below:

- Bat-fatality rates at wind turbines are variable across sites and regions of North America. Estimates of bat fatalities were highest at wind-energy facilities located on forested ridges in the eastern United States and the prairies of eastern Canada, and were lowest in the Rocky Mountain and Pacific Northwest regions.
- Bat-fatality rates range from just below one bat per installed megawatt per year (bats/MW/year) to as high as 70 bats/MW/year.

- Three species of migratory tree bats compose the majority of fatalities (hoary, eastern red, and silver-haired bats), and hoary bats compose about half of all documented fatalities in North America.
- Silver-haired bat fatalities were more frequently found in Canada, Iowa, and the Pacific Northwest relative to the eastern United States, whereas eastern red bats were commonly found in eastern forested sites.
- In Texas and Oklahoma, wind-energy development has the potential to negatively impact migratory Brazilian free-tailed bats (see Miller, 2008; Piorkowski and O'Connell, 2010).
- Fatalities were skewed toward males of the four most commonly killed species at most facilities.
- Most fatalities occur during a period of time that coincides with autumn migration, although few studies spanned the entire season when bats were active.
- Larger, taller turbines killed more bats per turbine when compared to smaller turbines. The highest bat-fatality rates occurred at turbines with towers of 65 m or taller.
- Bats collide with spinning turbine blades and do not strike stationary blades or towers.
- The majority of bats appeared to be killed on low wind-speed nights when power production was proportionally low but turbine blades were still moving, often at or close to fully operational speed.
- There appeared to be no difference in bat-fatality rates between turbines equipped with Federal Aviation Administration (FAA) lights and those that were unlit.
- Factors influencing fatality rates remain unclear, and it is not known yet how to identify and avoid high-risk sites for bats and wind energy.
- Population sizes of migratory tree bats are unknown because of their cryptic habits and limitations in how their numbers can be estimated. There is currently no established method of assessing how quickly the current rates of bat mortality at turbines could cause population declines.
- Studies in Canada and the United States have independently shown that curtailment, or preventing turbine blades from turning during relatively low wind speeds in late-summer and autumn, can reduce bat fatalities by as much as 40–90 percent.

II. Annotated Bibliography

The following is an annotated bibliography of 36 peer-reviewed publications and five unpublished theses in alphabetical order. For research papers, annotations include an overview of objectives, methods, key findings, and conclusions. For review or synthesis papers, the main highlights are provided.

Arnett, E.B., 2006, A preliminary evaluation on the use of dogs to recover bat fatalities at wind-energy facilities: Wildlife Society Bulletin, v. 34, p. 1440–1445.

This study was a baseline effort to assess the efficiency of dog-handler teams to recover bat fatalities at two wind-energy facilities in West Virginia and Pennsylvania. Objectives: (1) to train dogs to find bat carcasses; and, (2) to conduct pilot studies to determine the search efficiency of dog-handler teams under different vegetation conditions. Methods: The study took place at Mountaineer Wind Energy Center in West Virginia and the Meyersdale Wind Energy Center in Pennsylvania. The author trained two chocolate labs, one male and one female, for seven days prior to field testing. Dogs were trained by seeding a transect with bat carcasses representing different species at various stages of decay. Search efficiency of the dog-handler team was tested simultaneously with human searchers during scheduled trials at the two wind-energy facilities in late summer 2004.

Key Findings:

- Overall dog-handler efficiency (percent of trial bats found) for all trials and bats combined, and using combined findings from both dogs, was 71 percent at Mountaineer and 81 percent at Meyersdale, compared to 42 percent and 14 percent for humans, respectively.
- Quite a bit of variability was found in searcher efficiency (both dog-handler and humans) based on vegetative cover, terrain, and amount of high-visibility habitat found at the two sites.

Conclusions: Although the findings from this pilot study were promising, more research is warranted to better understand patterns and account for limitations and biases influencing the efficiency of dog-handler teams as a tool to improve fatality estimates for bats at wind-energy facilities.

Arnett, E.B., Inkley, D.B., Larkin, R.P., Manes, S., Manville, A.M., Mason, J.R., Morrison, M.L., Strickland, M.D., and Thresher, R., 2007, Impacts of wind-energy facilities on wildlife and wildlife habitat: Bethesda, Md., The Wildlife Society, Wildlife Society technical review 07-1,

This technical review summarizes information on the impacts of wind-energy facilities on wildlife and wildlife habitat, including state and federal permitting processes, wildlife fatality, habitat loss and modification, animal displacement and fragmentation, offshore development, and issues surrounding monitoring and research methodology, including the use of technological tools. In the review, a summary section is provided on bats and bat fatalities from wind turbines. This review provided 12 recommendations that should help managers and decision-makers meet the challenges of developing wind energy responsibly: (1) improve state agency involvement and consistency for requirements and regulation; (2) renewable portfolio standards (for example, regulations that require increased production of energy from renewable energy sources, such as wind, solar, biomass, or geothermal); (3) develop federal and state guidelines; (4) avoid siting wind facilities in high-risk areas; (5) reduce fragmentation and habitat effects; (6) conduct priority research; (7) evaluate pre-construction assessments and predicted impacts; (8) conduct more consistent, longer-term studies; (9) develop and evaluate habitat-related mitigation strategies; (10) employ principles of adaptive management; (11) conduct regional assessments and forecasting of cumulative land-use and impacts from energy development; and, (12) improve public education, information exchange, and participation.

Arnett, E.B., Brown, W.K., Erickson, W.P., Fiedler, J.K., Hamilton, B.L., Henry, T.H., Jain, A., Johnson, G.D., Kerns, J., Koford, R.R., Nicholson, C.P., O'Connell, T.J., Piorkowski, M.D., and Tankersley, R.D., Jr., 2008, Patterns of bat fatalities at wind-energy facilities in North America: Journal of Wildlife Management, v. 72, p. 61–78.

This synthesis examined the patterns of bat fatalities from post-construction studies at wind-energy facilities. Objectives: (1) present unifying and unique patterns from the studies; (2) discuss the scope, biases, and limitations of existing efforts; and, (3) identify information gaps and offer suggestions for future research needed to develop mitigation strategies to minimize or eliminate bat fatality at wind facilities. Methods: Authors reviewed 21 studies from 19 different wind-energy facilities in 5 regions of the United States and one province in Canada. These studies took place from 1996 to 2006 and represent only those published in peer-reviewed journals or unpublished reports available publicly.

Key Findings:

- The duration of most studies was short.
- Only eight studies used bat carcasses to quantify searcher bias and scavenger removal.
- Most studies calculated an overall bias correction factor for the entire study period and across different habitat conditions.
- Estimates of bat fatalities were highest at wind-energy facilities located on forested ridges in the eastern United States and lowest in the Rocky Mountain and Pacific Northwest regions.
- In most regions and individual studies, bat fatalities were heavily skewed to migratory foliage-roosting species including hoary bats (*Lasiurus cinereus*), eastern red bats (*L. borealis*), and migratory tree and cavity roosting silver-haired bats (*Lasionycteris noctivagans*).
- Hoary bats constituted high proportions of fatalities at most facilities.
- Silver-haired bats were found more frequently in Canada, Iowa, and the Pacific Northwest relative to the eastern United States, whereas eastern red bats were commonly found in eastern forested sites in the midwestern United States.
- Fatalities were skewed toward males for the four most commonly killed species at most facilities.
- The highest bat fatalities were consistently reported during late summer and early fall, although few studies spanned the entire season when bats are active (generally April–November).
- Evidence was found that larger, taller turbines killed more bats per turbine compared to smaller turbines.
- Findings from these studies support the contention that bats collide with spinning turbine blades and do not strike stationary blades or towers.
- The majority of bats appear to be killed on low wind nights when power production seemed insubstantial but turbine blades were still moving, often at or close to full operational speed.
- There appeared to be no statistically significant differences in fatality between turbines equipped with FAA lights and those that were unlit.

Conclusions: The authors end by addressing the limitations of the studies reviewed: (1) most work conducted was short term, and the frequency of the study was inadequate to assess seasonal and annual variability; (2) fatality estimates could not be directly compared among studies because of different sampling protocols; (3) sizes of search plots varied among studies; (4) search effort, intensity of searches, and experience of observers varied across studies; and, (5) most studies failed to adequately account for habitat variability during fatality searches.

Arnett, E.B., Huso, M.M.P., Schirmacher, M.R., and Hayes, J.P., 2011, Altering wind turbine speed reduces bat mortality at wind-energy facilities: Frontiers in Ecology and the Environment, v. 9, p. 209–214.

This paper describes an experiment to test the effectiveness of raising the turbine cut-in speed to reduce bat mortality at a wind facility in Pennsylvania during 2008 and 2009. Wind-turbine cut-in speed was defined to be the lowest wind speed at which turbines generate power to the utility system. Raising the wind-turbine cut-in speed renders turbines non-operational until the higher cut-in speed is reached; then turbines begin to spin and produce power. Objectives: (1) to determine if rates of bat fatality differed between fully operational turbines with a cut-in speed of 3.5–5.0 m s^{-1} and turbines with raised cut-in speeds of either 5.0 m s^{-1} and 6.5 m s^{-1}, and (2) to quantify the economic costs of different curtailment programs and timeframes. Methods: The study took place at the Casselman Wind Project in Somerset County, Pa., over a 2-yr period (2008–2009). Three turbine treatments were defined and assigned in a randomized block design to 12 turbines. Treatments were (1) fully operational, (2) cut-in speed of 5.0 m s^{-1}, and (3) cut-in speed of 6.5 m s^{-1}. Treatment-related mortality was measured as the sum of all individual carcasses of bats estimated to have been killed during the previous night observed along transects near the turbines.

Key Findings:

- No difference was found between number of fatalities for the two cut-in speed treatments for both years.
- Observed bat mortality at fully operational turbines was, on average, 5.4 (2008) and 3.6 (2009) times greater than mortality associated with curtailed (that is, non-operating) turbines.

Conclusions: Relatively small changes to wind-turbine operation resulted in nightly reductions in bat mortality, ranging from 44 percent to 93 percent, with marginal annual power loss (less than or equal to 1 percent of total annual output).

Baerwald, E.F., and Barclay, R.M.R., 2009, Geographic variation in activity and fatality of migratory bats at wind-energy facilities: Journal of Mammalogy, v. 90, p. 1341–1349.

This study examined geographic variation in activity levels and fatality rates of bats at one proposed and eight existing wind-energy facilities across southern Alberta, Canada. The authors hypothesized that migratory bats use specific migration routes associated with landmarks and roost availability. They predicted that migratory bat activity would vary among the different habitats and that bat fatalities at wind turbines would correlate with activity rate. Objectives: to determine if bat activity and fatality were concentrated in certain areas or evenly distributed across the landscape. Methods: Bat activity was monitored acoustically from July 15 to September 15, 2006 and 2007, at seven proposed or existing wind-energy installations. Anabat II detectors were installed in pairs (one approximately 1.5 m from the ground and the other 30 m from the ground) on meteorological towers. The authors compiled fatality data collected between 2001 and 2007 from nine wind-energy facilities. They also searched for bat carcasses at two sites from July 15 to September 30, in 2006, and 2007, and at another site from July 15 to September 30, 2007. Fatality rates were adjusted per turbine for searcher efficiency and scavenger removal rates.

Key Findings:

- Bat activity varied among sites with the highest activity of migratory-bat passes detected at Site 1, the westernmost site near the Rocky Mountains.
- Pooled across all sites, total migratory bat activity did not differ between detection heights. However, when separated by species, the activity of hoary bats and silver-haired bats differed between the two detection heights with hoary bats being less common at ground level than at 30 m and silver-haired bats more common at ground level than at 30 m.
- Searcher efficiency was estimated to be 78 percent and carcasses lasted an average of 5.6 days.

- Corrected fatality rates varied among the nine sites sampled, from 31.41 plus or minus 3.04 bats/turbine/year to 1.26 plus or minus 0.18 bats/turbine/year, with higher rates at western wind-energy facilities and lower rates at eastern facilities. Fatality rate increased with turbine height.
- Fatality rate was not influenced by migratory bat activity at ground level, migratory bat activity at 30 m, or the interaction between migratory bat activity at ground level and turbine height. Conclusions: Migratory bats in southern Alberta appear to be concentrating along select routes and activity rates of both hoary and silver-haired bats were higher near the foothills of the Rocky Mountains to the west than on the prairie grasslands farther east. Fatality rates of bats varied among all sites, partly due to differences in turbine tower height, but also due to the differences in activity of migratory bats at 30 m. The results of this study indicated that factors in addition to turbine height and migratory bat activity influence fatality rates at wind turbines. The authors suggested that the number of turbines may play a role: facilities with many turbines may experience a dilution effect with lower fatality rates per turbine than facilities with few turbines.

Baerwald, E.F., and Barclay, R.M.R., 2011, Patterns of activity and fatality of migratory bats at a wind energy facility in Alberta, Canada: Journal of Wildlife Management, v. 75, p. 1103–1114.

Using acoustic monitoring and carcass searches, the authors examined temporal and spatial variation in activity levels and fatality rates of bats at one wind-energy facility in southern Alberta, Canada, in 2006 and 2007. Objectives: to test the hypotheses regarding the influence of weather variables and turbine characteristics on migratory bat activity and fatality. The authors predicted that weather conditions conducive to migration would increase activity of migratory bats and therefore would increase fatality rate. They also predicted that weather and turbine characteristics would affect fatality rate independent of the number of bats migrating on a particular night (for example, bats attracted to turbines). Methods: Weather data were collected from a meteorological tower located on the wind energy facility and included wind speed and direction data at 50 m above ground level and temperature and barometric readings at 2 m above ground-level. Activity was monitored using Anabat II detectors on meteorological towers and turbine nacelles from July 15 to September 30, 2006 and 2007. Nacelles are the covers at the top of the turbine that house all of the generating components of a wind turbine, including the generator, gearbox, drive train, and brake assembly. Carcass searches were also conducted during the same time periods for both years and were adjusted for searcher efficiency and scavenger removal rates. Four different statistical analyses examined whether: (1) activity rate varied with weather; (2) variation in fatality rate was influenced by turbine location and characteristics; (3) variation in fatality rate was influenced by activity rate; and finally, (4) fatality rate depended on weather variables, while controlling for variation in activity of bats.

Key Findings:

- Activity of migratory bats was greater in low wind speeds and was lower when the wind was from the north or northeast. Hoary bats were more active with falling barometric pressure but were not influenced by any other weather variable. Silver-haired bat activity was greater in warmer ambient temperatures and low wind speeds and decreased when the wind was from the north or northeast.
- Fatality rates differed among turbines; higher rates were found at turbines in the north end of the facility in comparison to the south end. There were no differences in fatality rates at turbines on the east side versus the west side of the facility, at turbines on the end of a row versus in the middle of a row, or at turbines lit by aviation lighting versus those unlit.
- Bat fatalities increased with total migratory bat activity; however, when separated by species, hoary bat fatalities per night did not increase with their activity and silver-haired bat fatalities per night did increase with their increased activity.
- Total nightly fatalities increased with increased moon illumination and were influenced by the interaction between barometric pressure change and the activity of migratory bats. Although fatality rates increased as activity increased, if activity was low and there was a large drop in barometric pressure, the fatality rate was high. Hoary bat fatality rate increased with falling barometric pressure but not with any other variable or interaction tested. Silver-haired bat fatality rate increased with increasing activity, increasing fraction of moon illuminated, and southeasterly winds.

Conclusions: Migration behavior of bats is a very complex issue, but it offers potential ways to reduce fatalities at wind-energy facilities if we can understand the relationships between weather, activity, and fatality. As an example, if increasing moon illumination and falling barometric pressure increase the risk of fatality, then matching these parameters with changes to wind-turbine operations might reduce fatalities.

Baerwald, E.F., D'Amours, G.H., Klug, B.J., and Barclay, R.M.R., 2008, Barotrauma is a significant cause of bat mortalities at wind turbines: Current Biology, v. 18, p. 695–696.

The authors describe a study where they found that barotrauma was a significant cause of bat fatalities at wind turbines in southwestern Alberta, Canada. The decompression hypothesis, as described by Baerwald and others, 2008, proposes that bats are killed by barotrauma caused by rapid air-pressure reduction near moving turbine blades. Objective: to test the decompression hypothesis. Methods: Hoary bats and silver-haired bats killed at a wind facility were examined for external and internal injuries. Key Findings:

- Of 188 bats killed at turbines the previous night, 87 had no external injuries that would have been fatal.
- Of 75 fresh bats necropsied, 32 had obvious external injuries, but 69 had hemorrhaging in the thoracic and/or abdominal cavities.

Conclusions: Barotrauma helps explain the high fatality rates of bats at some wind-energy facilities. Even when echolocation allows bats to avoid turbine blades, they may still be incapacitated or killed by the internal injuries caused by rapid pressure reductions.

Baerwald, E.F., Edworthy, J., Holder, M., and Barclay, R.M.R., 2009, A large-scale mitigation experiment to reduce bat fatalities at wind-energy facilities: Journal of Wildlife Management, v. 73, p. 1077–1081.

This paper describes the results of a large-scale mitigation experiment to reduce bat fatalities at a wind facility in southwestern Alberta, Canada, during the peak period of migration for hoary and silver-haired bats in 2006 and 2007. Normal operating turbines at this facility had a cut-in speed (the wind speed at which the turbine begins to generate electricity) of 4 m/s. Below that wind speed, the turbine rotor rotates at a slow rate that increases with wind speed until the rotor is turning at a rate which triggers generator rotation, a wind speed of 4 m/s. Objective: to test the hypothesis that reducing the amount that turbine rotors turn in low wind speeds reduces bat fatalities. Methods: Twenty-one randomly chosen turbines were operationally altered in two ways in 2007. Fifteen had their rotor startup speed increased to 5.5 m/s, meaning that the turbines were idle and motionless during low wind speeds. The other six turbines were manipulated to change the pitch angle of their blades to reduce rotor speed in low wind, thereby lowering the generator speed required to start energy production, which caused turbines to be motionless. This procedure was similar to the other experimental treatment but had different implications for turbine operations. Fatality searches were conducted once per week using transects and two searchers. Corrected fatality rates were compared between experimental and control turbines both in 2006 (when no experiment was conducted) and in 2007.

Key Findings:

- In 2006, there was no difference in corrected bat-fatality rates between turbines later selected as experimental or control.
- In 2007, both sets of experimental turbines killed fewer bats than did control turbines, and there was no difference between the two experimental treatments.
- Corrected fatality rates for each species of migratory bat were reduced by between 50 percent and 70 percent at the experimental turbines.
- By increasing the rotor start-up wind speed, the amount of time the turbines produced electricity was reduced by an average of 42.3 percent.

Conclusions: The two experimental changes made to operation of turbines had a similar effect on their operation at low wind speeds and therefore resulted in lower bat fatalities. However, the effect of changes in operation was different in terms of costs with increased rotor start-up speed (5.5 m/s) being more costly than changing the pitch angle of the blades.

Barclay, R.M.R., Baerwald, E.F., and Gruver, J.C., 2007, Variation in bat and bird fatalities at wind-energy facilities: assessing the effects of rotor size and tower height: Canadian Journal of Zoology, v. 85, p. 381–387.

The size and height of wind turbines have increased over the years, from 18-m diameter rotors on top of 24-m towers, to the newest turbines with 90-m diameter rotors and towers as tall as 94 m. Objective: to test the hypothesis that wind turbine size and height influence fatality rates of bats and birds and to determine whether the effect differs between the two groups. Methods: Authors compiled data on the characteristics of 33 North American wind-energy facilities and on bat and bird fatalities. Data were available from some published scientific papers, but primarily from unpublished government, industry, and consultant reports, and from personal communications. Fatality rates were determined based on searches for carcasses under turbines.

Key Findings:

- Increased rotor-swept area was not a significant factor in fatality of birds or bats.
- Fatality rates of both bats and birds were relatively low at short turbines (less than 65 m high), but bat fatalities increased exponentially with turbine height, while bird fatalities did not change.
- The highest bat-fatality rates occurred at turbines with towers 65 m or taller. Conclusions: Increased rotor diameter and tower height of turbines may result in greater energy output per turbine, but these larger turbines have higher bat-fatality rates than bird-fatality rates.

Cryan, P.M., 2008, Mating behavior as a possible cause of bat fatalities at wind turbines: Journal of Wildlife Management, v. 72, p. 845–849.

This commentary proposes a hypothesis that tree bats collide with turbines while engaging in mating behaviors that center on the tallest trees in a landscape. The author further proposes that these behaviors occur because of two different mating systems: resource defense polygyny and lekking. Key points: Tree bats may use the highest trees in a landscape as rendezvous points during the mating period, and this behavior places them at risk of collision with the blades of wind turbines. Reproductive bats could be attracted to the turbines when looking for mating opportunities and mistake the turbines for the tallest trees. Male European tree bats have been observed defending roosting structures needed by females during the mating season (late summer and autumn). This mating strategy is called resource defense polygyny. However, tree bats in North America (species of *Lasiurus* or lasiurines) differ in roosting habits and behavior from European species because they roost alone or in small family groups in tree foliage during spring and early summer. Foliage roosts used by female lasiurines during fall migration are most likely abundant and not easily defended by male bats. Lekking is a mating system for species in which females range widely or resources (food or roost) are too finely distributed for males to defend. Lekking could explain why so many lasiurines are recovered in greater number and more frequently than any other species of bats killed by wind turbines in North America. An interesting characteristic of lekking that makes this high casualty rate among lasiurines plausible is that males tend to create dense aggregations during late summer and autumn to sites visited by females for mating. Conclusions: Research focused on hypotheses of attraction should be prioritized. If courtship and mating activities occur at wind turbines, turbines have the potential to create population sinks that attract bats from far away, and selectively kill bats that are primed for reproduction.

Cryan, P.M., 2011, Wind turbines as landscape impediments to the migratory connectivity of bats: Environmental Law, v. 41, p. 355–370.

This review summarizes bat mortality caused by wind turbines and the potentially significant effects of this mortality on certain bat populations. The article includes an overview of bat migration, discusses natural and human-caused mortality of migrating bats, provides background on bat fatalities at wind turbines, and highlights the importance of future innovative research.

Key points:

- Many species of temperate-zone bats migrate long distances; however, the details of their migratory movements and seasonal whereabouts are mostly lacking. This information is lacking for several reasons: (1) bats migrate at night; (2) they are inactive during the day; (3) they roost in concealed and frequently inaccessible locations; and, (4) there have been few recorded observations of actively migrating bats.
- Migration in bats most likely evolved in response to their distinctive thermoregulatory strategies and life histories. They mate during autumn and winter, and females often migrate in spring while they are pregnant, with some moving to habitats and regions of the continent separate from males during the early summer. Many species of bats can also drop their metabolism throughout the year and enter torpor to save energy when climate, weather, or food availability is unfavorable. Bats also have high rates of adult survival, resulting in longevity. Reproductive success is also high due to their putting considerable time and energy into taking care of a small number of young. These life-history strategies result in bat populations growing

slowly and possessing an inability to quickly rebound after rapid declines in population size.

- Migration can be a perilous endeavor. Migrating bats may experience higher natural mortality than species that do not migrate.
- Collisions of migrating bats with buildings, communication towers, and aircraft have not been an obvious conservation issue in the past.
- Wind turbines are an emerging threat to bat migrations.
- Bat-fatality rates at wind turbines are variable across sites and regions in the United States. Fatality rates range from just below one bat per installed megawatt per year to as high as 70 bats per megawatt per year. Based on published fatality rates and the number of currently installed wind turbines, an estimated 450,000 bats may already perish at turbines each year in North America.
- Three species of migratory tree bats compose the majority of fatalities, and hoary bats compose about half of all documented fatalities in North America.
- Factors influencing fatality rates remain unclear, and studies at sites in North America have not found any consistent relationships between local landscape features and fatality rates, nor do we have a clear understanding of the regions or types of habitats where fatalities most likely occur. Therefore, it is not known yet how to identify and avoid high-risk sites for bats and wind energy. Most fatalities occur during a period of time that coincides with mating and autumn migration.
- Population sizes of migratory tree bats are unknown because of their cryptic habits and limitations in how their numbers can be estimated. Therefore, there is no way of assessing how quickly the current rates of bat fatality at turbines could cause catastrophic population declines and possible species extinctions. However, more hoary bats have been estimated to die at certain wind-energy facilities in the United States in two to three years than have ever been collected and preserved as scientific study specimens in the museums of the Americas.
- Research on the proximate causes for bat fatalities at wind turbines is essential to help gauge the gravity of the situation and the need for conservation measures. Turbines may act as barriers to migration and disrupt necessary migratory connectivity. They may also act as population sinks and bottlenecks. Without knowing the exact reasons why bats are susceptible to turbines, scientists trying to find ways to minimize fatalities will continue to grope for solutions.

- None of the bat species most affected by wind energy are protected by federal conservation laws such as the Endangered Species Act (ESA). Neither are these bats protected by legislation pursuant to international treaties such as the Migratory Bird Treaty Act.
- Studies in Canada and the United States have independently shown that curtailment, or preventing turbine blades from turning during relatively low wind speeds in late-summer and autumn, can reduce bat fatalities by as much as 40 to 90 percent over the course of the studies.

Conclusion: The author concludes that a legal framework for proactively monitoring and protecting all migratory wildlife before their populations become endangered may be the type of foresight needed to help drive scientific advances. This may also allow us to better predict and deal with emerging threats to migratory wildlife.

Cryan, P.M., and Barclay, R.M.R., 2009, Causes of bat fatalities at wind turbines—Hypotheses and predictions: Journal of Mammalogy, v. 90, p. 1330–1340.

This synthesis reviews the hypothesized causes of bat fatalities at wind turbines. The hypotheses of cause fall into two categories: proximate and ultimate. Proximate causes explain the direct means by which bats die at turbines and include bats colliding with turbine towers, colliding with rotating blades, or suffering internal injuries after being exposed to rapid pressure changes near the trailing edges and tips of the moving blades (barotrauma). Hypotheses of ultimate causes are numerous and include three general categories: random collisions, coincidental collisions, and collisions that result because bats are attracted to turbines. Random collisions are those that occur due to chance alone or no assumptions of circumstance or attraction. Coincidental collisions involve bats being victims of unfortunate behavioral circumstances that put them at risk of colliding with turbines (for example, turbines are located along migratory pathways). Hypotheses of attraction propose that there is some attractor or combination of attractors drawing bats to wind turbines. Table 1 of this publication provides an overview of the hypotheses and their corresponding predictions. The authors conclude that the conservation implications of bat attraction to wind turbines are by far the most serious of the causes and suggest that testing the general hypotheses of attraction be a top research priority.

Cryan, P.M., and Brown, A.C., 2007, Migration of bats past a remote island offers clues toward the problem of bat fatalities at wind turbines: Biological Conservation, v. 139, p. 1–11.

Little is known about the behavior of bats during migration and what aspects of this seasonal behavior make them particularly susceptible to mortality from wind turbines. Objective: to analyze a 38- yr data set of observations on hoary bats on Southeast Farallon Island to determine if moonlight and weather conditions influenced the occurrence of bats on the island. The authors also examined the temporal relationships of migration events involving hoary bats and similar species, and then related those findings to observed bat fatalities at wind turbines. Methods: This study took place at Southeast Farallon Island (SEFI), part of the Farallon National Wildlife Refuge. Personnel at the Point Reyes Bird Observatory recorded daily observations from routine censuses of birds and marine mammals and recorded regular observations of hoary bats from as early as 1968. Logistic regression was used to model the influence of weather and moonlight on bat arrival to the island and bat departure from the island. To summarize the seasonal timing of known events involving migration by species of *Lasiurus*, the authors compiled records from the literature.

Key Findings:

- Bats were observed more frequently in September than any other month and usually visited the island between the last week of August and the end of October.
- For describing the arrivals of bats, the model that included wind speed, barometric pressure, cloud cover, and moon illumination ranked the best.
- The top-ranked model describing departures included only wind speed, cloud cover, and moonlight.
- Low wind speeds, moon illumination, and relatively high degrees of cloud cover were important predictors of both bat arrivals and departures, whereas low barometric pressure was an additional condition that helped predict arrivals.
- The majority of records compiled from the literature occurred on days between early September and late October, with a peak in mid-September.

Conclusions: The results from this study show that bats were more likely to arrive and depart from the island on evenings when wind speeds were relatively low. There was a predominance of hoary bat arrivals and departures on SEFI during darker phases of the moon and during overcast nights. There were five reasons this study's findings are relevant to the problem of bat fatalities at wind turbines: (1) arrivals of migrating tree bats at particular sites may be predictable events; (2) there is conclusive evidence that tree bats migrate over offshore waters, on which future construction of wind turbines has been proposed; (3) migratory hoary bats use vision to navigate during migration and are drawn toward visual stimuli; (4) there is a clear relationship between the timing of autumn migration by species of *Lasiurus* and collisions with anthropogenic structures; and, (5) there is a possible behavioral explanation for bat collisions with wind turbines.

Fiedler, J.K., 2004, Assessment of bat mortality and activity at Buffalo Mountain Windfarm, eastern Tennessee: Knoxville, Tenn., University of Tennessee, Unpublished Master's Thesis. 166 pp.

This thesis describes an investigation of bat mortality at Tennessee Valley Authority's Buffalo Mountain Windfarm (BMW) in eastern Tennessee. Objectives: (1) to document the extent of bat mortality at BMW and then examine the mortality for spatial, temporal, and species patterns; (2) to determine bat activity using bat detectors and mist netting; and, (3) to determine the relationships between bat activity and mortality. Methods: Mortality monitoring was conducted during a three-year period beginning late September 2000 and concluded the end of September 2003. Carcass searches were conducted twice weekly from April 1 through November 1. Logistic regression was used to identify possible explanatory variables to predict nights when a fatality event would likely occur. Explanatory variables included daily generation of the wind facility, average wind speed, average wind direction, average temperature, minimum and maximum temperature between 1800 and 0600 hours, and the difference between minimum and maximum temperature. Biases in searcher efficiency, proximity and density of scavengers in the area, and changes in vegetation structure were accounted for by conducting six search bias trials. Bat activity data were collected with mist-netting and acoustic monitoring (Anabat II detectors).

Key Findings:

- Over the course of the study, a total of 120 bat fatalities of six species were found.
- Seventy percent of bat fatalities occurred from August 1–September 15.
- An adjusted mortality rate of 20.82 bats/turbine/year was calculated for BMW.
- Eastern red bats (*Lasiurus borealis*) made up the majority of bat fatalities overall at 61.3 percent. Tricolored bats (*Perimyotis subflavus*) and hoary bats were the next two most common fatalities overall (22.4 percent and 10.1 percent, respectively). The remaining bat fatalities consisted of three species: big brown bats (*Eptesicus fuscus*), silver-haired bats, and one Seminole bat (*Lasiurus seminolus*).
- More males than females were killed.
- The peaks in bat mortality coincided with months of low winds and less wind-energy-facility generation.
- A fatality occurrence was more likely during slightly cooler nights with calmer, less variable winds.
- Searcher detection rates for carcasses was highly variable and ranged from 9 to 75 percent for the six trials.
- The average number of days a scavenged carcass remained onsite for the six trials ranged from 3.2 to 16.7 days.
- Thirteen possible scavenger species were identified, and 36 percent of planted carcasses were removed from the site before the search the next morning.
- Acoustical monitoring of bat activity showed that annual bat-activity levels were consistently greater during late summer and early fall than spring and summer.
- Bat activity near water was greater than away from water.
- On nights where fresh fatalities occurred, bat-activity levels were greater than on nights where no fatalities occurred.
- Turbine collisions of red bats and hoary bats were disproportionately greater than acoustical monitoring indicated at BMW.

Conclusions: Several biases may affect estimates of bat mortality rates at wind facilities. These include searcher efficiency, scavenging rates, search interval, background mortality rates, crippling bias, plot size, and seasonal vegetation cover. The author recommends that individual wind-energy facilities estimate these biases and their precision when calculating mortality rates, as both regional and site-specific variables can influence them.

Grodsky, S.M., 2010, Aspects of bird and bat mortality at a wind facility in southeastern Wisconsin— Impacts, relationships, and cause of death: Madison, Wis., University of Wisconsin, Unpublished Master's Thesis. 99 pp.

This thesis presents results of a post-construction study of bird and bat mortality at wind turbines at the Forward Energy Center in southeastern Wisconsin. The thesis consists of two chapters both intended to be submitted for publication in peer-reviewed journals. The first chapter describes a study assessing bird and bat mortality at a wind energy facility in southeastern Wisconsin. The second chapter investigates the cause of death for wind turbine-associated bat fatalities. The results of the second chapter are summarized in the publication following this thesis (see below), so this summary will describe the results of the first chapter only. Objectives: (1) to determine estimated mortality rates for birds and bats, both migratory and non-migratory; (2) to examine species composition of wildlife mortality; and, (3) to correlate variables associated with bird and bat mortality. Methods: This study took place in the fall from July 15–October 15, 2008 and 2009, and the spring from April 15–May 31, 2009 and 2010. Twenty-nine wind turbines were randomly selected and searched for fatalities using a stratified sampling technique. Searcher efficiency trials were conducted to estimate percentage of carcasses not seen and recovered by individual searchers. Scavenger removal rates were estimated using removal trials. A mortality estimator was used that accounted for searcher efficiency, scavenger removal, search interval, density-weighted area searched, and visibility within search transects. Covariates included in the analysis were weather variables, turbine operating status, bat activity data from Anabat detectors, and distances to a nearby marsh and mine. A generalized linear mixed model was used to look at the relationship between bird and bat mortality and covariates.

Key Findings:

- A total of 20 birds were recorded during carcass searches, and the estimated bird-mortality rates averaged across all study periods were 0.0255 birds/turbine/day and 0.0170 birds/MW/day.

- Most bird species recorded as fatalities were nocturnally migrating passerines.
- A total of 122 bat carcasses were found (of five species).
- More than half of the bat mortalities were hoary bats and silver-haired bats. When eastern red bat carcasses were included, the migratory tree bats accounted for 69 percent of the bat mortality.
- Most bats were adults, and females outnumbered males.
- Averaged across all study periods, migratory bat mortality rates were 0.158 bats/turbine/day and 0.105 bats/MW/day. Non-migratory bat mortality rates were 0.0382 bats/turbine/day and 0.0255 bats/MW/day.
- Most bat mortality occurred from late August through the second week of September.
- Modeling suggested that temperature affected bat mortality significantly and distance to the marsh affected bat mortality marginally. All other covariates modeled did not influence bat mortality.
- Searcher detection rate was 38 percent for bats and 63 percent for birds. Conclusions: The underlying trend to all current mortality studies in the Midwest is that there is significantly higher bat mortality rates compared to bird mortality rates.

Grodsky, S.M., Behr, M.J., Gendler, A., Drake, D., Dieterle, B.D., Rudd, R.J., and Walrath, N.L., 2011, Investigating the causes of death for wind turbine-associated bat fatalities: Journal of Mammalogy, v. 92, p. 917–925.

This study investigated the causes of death for wind-turbine-associated bat fatalities at the Forward Energy Center in southeastern Wisconsin. Objectives: (1) to use veterinary diagnostic procedures to determine the cause of lesions found on bats killed by wind turbines; (2) to correlate patterns of injuries with causes of death; and, (3) to create guidelines for future studies. Methods: Bat carcasses were collected July 15–October 15, 2009, using techniques described above in Grodsky (2010). Thirty-nine bat carcasses were radiographed and evaluated by a board-certified veterinary radiologist who was unaware of the postmortem findings. Data recorded included the location, number, and type (that is, fracture classification, dislocations) of skeletal injuries and any abdominal or diaphragmatic hernias. The thoracic cavity was examined for radiographic signs of pneumothorax (air in the thoracic cavity), hemothorax (blood in the thoracic cavity), chest wall trauma, atelectasis, or pleural fluid accumulation. Eighteen bat carcasses were also dropped off the top of the turbine nacelle 91.44 m above ground level to see whether bats could break bones from a free fall. These dropped bats were then radiographed and the same veterinary radiologist reviewed the radiographs. Bats were also necropsied, and histologic sections of the lung and middle ear tissues were made. All brain samples were also tested for rabies.

Key Findings:

- Of the 39 bats radiographed, at least one broken bone was detected in 74 percent of the bats. Fractures to the wing bones were the most common, and the majority of the fractures were comminuted (that is, shattered or crushed into many pieces). Fractures to the skull, scapula, lumbar vertebrae, ribs/sternum, and pelvis were the next most common injuries.
- Of the 18 bats dropped, no new bones were fractured in most of the bats postdrop (80 percent).
- Of 33 bats necropsied, the most common lesions found were pneumothorax and hemothorax. An equal mix of inguinal and diaphragmatic hernias was the next most common injury.
- Of the 28 bats processed for histopathology, 5 (21 percent) had no pulmonary hemorrhage, 3 (13 percent) had mild hemorrhages, and 8 (33 percent) had moderate and severe hemorrhages.
- Middle ears from 24 bats were sampled and 11 (48 percent) bats had no hemorrhage in middle ears or surrounding structures, 4 (17 percent) bats had middle ear hemorrhage, and 8 (35 percent) bats had hemorrhage in and around the middle or inner ears.
- All bats tested negative for rabies.
- Thirty-eight bat-carcass locations were measured relative to distance from base of the turbine. The majority of the bats (71 percent) were found within 30 m of the base of the turbine. The highest percentage of bats (29 percent) was found within 10 m of the base of the turbine.

Conclusions: The exact cause of death (that is, barotrauma or direct collision) could not be determined in most bats due to the variability of injuries and a lack of exclusively attributable lesions. The authors suggest that the cause of death for bats killed at wind turbines is not exclusively or predominantly barotrauma or direct collision but rather an indiscernible combination of both. Visually inspecting bat carcasses is not adequate for conclusively diagnosing fatal injuries, including broken bones. The authors conclude that there may be delayed lethal effects of wind turbines on bat mortality. Bats could become crippled by either inner ear damage or by bone fractures, but survive long enough to exit the search area of mortality studies.

Gruver, J.C., 2002, Assessment of bat community structure and roosting habitat preferences for the hoary bat (*Lasiurus cinereus*) near Foote Creek Rim, Wyoming: Laramie, Wyo., University of Wyoming, Unpublished Master's Thesis. 164 pp.

This thesis examined bat community structure and roosting habitat preferences for the hoary bat (*Lasiurus cinereus*) proximate to the Foote Creek Rim Wind Power Facility in south-central Wyoming. Objectives: (1) assess bat-community composition proximate to the wind-energy facility; and, (2) quantify roosting habitat for the hoary bat. Methods: Mist-netting surveys were conducted over streams and ponds, and acoustic surveys (Anabat II) were conducted on rim-top, over ponds surrounding the rim, and over streams and ponds on the Medicine Bow National Forest. Searches for carcasses were conducted on transects within 30×30-m plots centered on selected turbines on Foote Creek Rim. Proportions of bats recovered during carcass searches in 2000 and 2001 were compared to those captured in mist-nets during the same period using a Chi-square test. Radio-telemetry was used to track hoary bats to daytime roosts, and habitat surrounding known roosts was quantified and compared to habitat surrounding randomly selected sites.

Key Findings:

- Bats in the genus *Myotis* represented approximately 81 percent of all bats captured (*Myotis volans* and *M. lucifugus*) in mist nets.
- The hoary bat was the fourth most abundant species captured with mist nets.
- Bat activity on the rim was exceedingly low compared to sites off the rim.
- Acoustic surveys revealed similar proportions of each group of bats found in mist-netting surveys, indicating that the assemblage of bats in the study area was comprised primarily of *Myotis* species (*M. lucifugus* and *M. volans*).
- Carcass searches in 2001 found 22 hoary bats and one big brown bat (*Eptesicus fuscus*).

- Over 3 yr of searches, hoary bats represented the overwhelming majority of recovered bats.
- Bat mortality at wind turbines by species was disproportionate to the relative abundance determined by mist netting.
- Radio transmitters were attached to 14 adult hoary bats, 6 of which were never located. Eight individuals were tracked to 17 different roosts. Instrumented bats roosted primarily in live lodgepole pine trees, although two individuals were found in medium-sized understory deciduous shrubs.
- Hoary bats preferred trees that were taller, had more canopy cover, and were near water. Conclusions: The author concludes this thesis by summarizing bat-turbine interactions. Deaths of bats at wind-energy facilities at the time of this thesis had only recently been reported and addressed. Records of turbine-related bat mortality were steadily increasing in the United States, and the cumulative results suggested two noteworthy trends: (1) most deaths are detected from late summer to early fall; and (2) bats in the genus *Lasiurus* (specifically *L. borealis* and *L. cinereus*) appear to represent the majority of recovered carcasses.

Horn, J., Arnett, E.B., and Kunz, T.H., 2008, Behavioral responses of bats to operating wind turbines: Journal of Wildlife Management, v. 72, p. 123–132.

This study directly addresses the causal factors of why bat fatalities occur at wind-energy facilities. The authors carried out this study by using thermal infrared (TIR) cameras to monitor the activity of bats at the Mountaineer Wind Energy Center in West Virginia. Objectives: (1) to document how bats behaved while flying within the rotor-swept zone where there was potential for direct contact with turbine rotors; (2) to examine temporal patterns of activity and to determine if environmental variables such as wind speed and temperature were associated with variation in bat activity levels; and, (3) to compare activity levels between lighted and unlighted turbines at the facility and to test if aviation obstruction lighting mounted on turbines attracts bats (either directly, or indirectly through increased insect abundance). Methods: Three TIR cameras were mounted on tripods and were placed 0.5 m apart to form a single observation point. Real-time streams of radiometric data from each of the three cameras was captured using software and laptop computers. This observation station was placed near the base of wind turbines at dusk and was positioned as close to 30 m as possible from the base of each turbine. Three 9-hr video sequences were recorded nightly for 10 nights. To test the potential of aviation lighting on the turbines to attract bats, cameras were placed alternatively at randomly selected lighted and unlighted turbines for 5 nights each. Data were analyzed by manually observing playback of all video sequences in real time and recording the appearance and timing of flying objects.

Key Findings:

- For all cameras combined, 4,568 moving objects were observed: 1,810 were bats (39 percent), 872 insects (19 percent), 46 birds (1.0 percent), 5 aircraft (0.1 percent), and 1,835 unknown (40 percent).
- Most bat activity near wind turbines occurred in the first 2 hr after sunset.
- Among the flying bats, the altitude above ground level (AGL) was highly variable, with some individuals flying within 10 m AGL and others foraging at or above the height of the turbine nacelle (70 m AGL).
- The vast majority of bats observed were flying at the medium-altitude band (within the rotor-swept zone; 65.0 bats/night). This was greater than three times the number observed flying at low or high altitudes (22.7 and 11.2 bats/night, respectively).
- Aviation lighting did not appear to affect the incidence of foraging bats around turbines.
- Mean turbine RPM and insect abundance were the most significant predictors of the number of bat passes observed.
- Clear instances of bats avoiding turbine blades and of bats being directly struck by blades were recorded. From 998 total bat observations, avoidance was observed 41 times (4.1 percent). Direct contact with blades was seen 5 times (0.5 percent).
- Exploratory behavior of bats was observed; bats often made several check passes or made repeated flight loops near the moving blades.

Conclusions: The rate of mortality at the Mountaineer facility suggests that bats are more abundant near turbines, and this study provides evidence that collisions may be nonrandom interactions between bats and moving turbine blades. One of the most important observations from this study was that bats were observed actively investigating both moving and motionless turbine blades; this suggests that the bats may be attracted to the turbines as potential roosting trees. The authors suggest that this study has ramifications for mitigating bat fatalities at wind facilities. For example, curtailment operations could occur during predictable nights or periods of high bat kills and could reduce fatalities considerably, with potentially modest reduction in power production and associated economic impact on project operations.

Huso, M.M.P., 2010, An estimator of wildlife fatality from observed carcasses: Environmetrics, v. 22, p. 318–329.

This research article addresses the problems of estimating numbers of wildlife fatalities from observed carcasses. Objectives: (1) formalize the conceptual model of fatality and define relevant parameters to provide a unifying framework for discussion; (2) develop an estimator of fatality when probability of detection is less than 1; (3) evaluate the bias and precision of the estimator using simulation and examine its sensitivity to magnitude of parameters as well as to assumptions regarding distributions of parameters; and, (4) compare the bias and precision of this estimator with two other estimators in common use. Methods: Simulations were conducted and included four major components: (1) simulate the fatality process; (2) simulate the carcass persistence (CP) using three distributional assumptions; (3) simulate the search process, using six search intervals; and, (4) simulate the observation process using two distributional assumptions to produce simulated carcass counts. Recorded bat echolocation calls at proposed turbine locations in Casselman, Pennsylvania were used in the simulation.

Key Findings:

- The proposed fatality estimator accounts for searcher efficiency, scavenger removal, search interval, density-weighted area searched, and visibility within search transects.
- None of the three estimators compared were unbiased under all conditions.
- Bias in the proposed estimator never exceeded plus or minus 27 percent, whereas bias in the other two estimators was always negative and exceeded that of the proposed estimator in 98 percent and 93 percent of the simulated conditions, respectively.
- The proposed estimator was relatively robust to variation in sources and magnitudes of imperfect detectability, but was sensitive to distributional assumptions regarding carcass removal rates and searcher efficiency.

- The proposed estimator offers significant improvement over two current estimators and provides relatively unbiased estimates of fatality that can be applied under a variety of conditions and survey protocols.

Jain, A.A., Koford, R.R., Hancock, A.W., and Zenner, G.G., 2011, Bat mortality and activity at a northern Iowa wind resource area: American Midland Naturalist, v. 165, p. 185–200.

This study examined bat collision mortality, activity, and species composition at an 89-turbine wind-resource area in farmland of north-central Iowa from mid-April to mid-December, 2003, and mid-March to mid-December, 2004. Objectives: (1) estimate the number of bat fatalities due to turbine collisions; (2) compare bat echolocation activity between areas with wind turbines and adjacent crop fields; and, (3) describe any trends in bat activity and mortality related to seasonal, weather, and habitat-related factors. Methods: The study took place at Top of Iowa Wind Resource Area (WRA). Twenty-six of 89 wind turbines were searched for evidence of collision-induced mortality. A stratified, randomized design was used to sample turbines. Three-m wide, 76 m-long transects under the turbines were kept free of vegetation by using herbicides and manual weeding. The average total area searched under each turbine was 1,742 square meters (m^2), which was about 30 percent of the 5,776 m^2 search plot. Searches occurred every 2 to 3 days. Fatality estimates were adjusted for search probability, carcass retention while exposed to scavengers, search efficiency, and proportion of turbines searched. Search efficiency and scavenge rates were estimated using small birds, since bat carcasses were not available. Mortality was compared between turbines with three different types of FAA lighting. Bat activity was monitored with two Anabat II detectors at 14 of the 26 turbines. One detector was placed at the base of the turbine and another was placed at the edge of an adjacent field. Total passes recorded at turbines versus adjacent sites were analyzed using a paired t-test. Geographic Information System (GIS) was used to estimate the distance of the nearest woodlot to each turbine monitored. Bat activity at turbines with three different types of FAA lighting was also compared.

Key Findings:

- In 2003, 30 bat carcasses of 5 species were found (11 hoary, 9 little brown, 5 eastern red, 3 big brown, and 2 silver-haired bats). In 2004, 45 bats of 6 species were found (11 hoary, 13 eastern red, 8 little brown, 5 big brown, 7 silver-haired, and 1 eastern pipistrelle *sic* tri-colored bat).

- All bats were found from June to October in 2003 and from June to September in 2004.
- Eighty-three percent of the bat carcasses in 2003 and 67 percent in 2004 showed obvious signs of trauma with bone fractures and hemorrhaging.
- Two of the 19 carcasses in 2003 had internal thoracic hemorrhaging without other evidence of physical trauma (barotrauma), but there was no evidence of barotrauma in 2004.
- Adjusted estimates of bat mortality of Top of Iowa WRA were 396.05 plus or minus 71.71 (95 percent CI) bats (4.45 bats/turbine, 4.94 bats/MW) between April, 2003, and December 15, 2003, and 635.83 plus or minus 111.89 (95 percent CI) bats (7.14 bats/turbine, 7.94 bats/MW) between March 24, 2004, and December 15, 2004.
- No significant difference in bat mortality was observed between turbines with the three different types of FAA lighting.
- There was no significant difference between bat activity at turbine sites and adjacent crop fields without turbines in 2003 and 2004.
- There was no significant effect of distance to nearest woodlot on the number of bat calls recorded at turbines.
- No difference was found in bat activity at turbines with red blinking and non-blinking beacons. Conclusions: The estimated mortality at this wind facility was higher than those of other wind facilities in the Midwest. Several factors were discussed that may cause bats to be susceptible to collision with wind turbines. These included local and regional habitat, population levels of bats, flight heights, weather and visibility conditions, turbine lighting, and use of the regions immediately around the turbines.

Jameson, J.W., 2010, Tall structure attraction and mortality at wind turbines of migratory tree bats in Manitoba Canada: Winnipeg, Manitoba, Canada, The University of Winnipeg, Unpublished Master's Thesis. 135 pp.

This thesis consists of four chapters: a general introduction, a chapter on testing the reproductive landmarks hypothesis, a chapter on factors influencing bat mortality at a wind-energy facility in central Canada, and a concluding chapter with recommendations. This summary focuses on Chapters 2 and 3.

Chapter 2 tests the reproductive landmarks hypothesis to explain mortality of bats at wind turbines. The "Reproductive Landmarks Hypothesis (RLH)" posits that bats may be attracted to tall structures during migration and that this attraction serves in finding mates. Objectives: (1) to confirm the attraction of migratory tree bats to turbines and to other tall anthropogenic structures; and, (2) to determine whether this attraction is associated with migratory and/or reproductive behavior. Methods: This study took place during the summer and fall of 2008 and 2009 at the St. Leon Wind Energy Project in south-central Manitoba, Canada. Bat activity was sampled at four groups of three habitat types: telecommunication service towers, open fields, and woodlots. Acoustic activity was monitored using Pettersson D240X time expansion bat detectors. Detectors were placed at 30 m on the towers. Generalized linear models were used to compare bat activity among habitat types and sites. Feeding activity (feeding buzzes) was also quantified and compared among habitat types using generalized linear models. Bat activity was also compared between ground level and 30 m communication towers. Mortality surveys were conducted at five wind turbines between July 6 and September 27, and bat mortality was correlated with bat activity at ground level and 30 m at communication towers. Carcass surveys were also conducted at communication towers.

Key Findings:

- In 2008, 4.8 percent of total bat passes were collected at open fields, 54.1 percent at woodlots, and 41.0 percent at communication towers.
- In 2008, 64.0 percent of passes were identified as silver-haired bats, 8.7 percent were hoary bats, 19.0 percent were eastern red bats, 7.2 percent were big brown bats, and 1.1 percent were little brown bats.
- In 2009, 9.6 percent of total bat passes were collected open fields, 45.1 percent at woodlots, 37.6 percent at communication towers, and 7.7 percent at wind turbines.
- In 2009, 64.1 percent of passes were identified as silver-haired bats, 24.1 percent were hoary bats, 8.1 percent were eastern red bats, 3.2 percent were big brown bats, and 0.65 percent were little brown bats.
- High levels of bat activity were recorded at communication towers during migration.
- Activity of silver-haired bats at communication towers during the pre-migration period was below activity levels at woodlots and equal to activity levels at open habitat—this is consistent with the second prediction of the RLH, that activity at towers would be relatively low

during pre-migration, as it is at wind turbines. Hoary bat activity was also consistent with this prediction.
- There were fewer tree-bat passes and less variability in activity at turbines when compared to communication towers.
- Equal proportions of feeding buzzes were observed across all habitats—this does not support the hypothesis that tall structure attraction in bats is linked to foraging behavior.

Chapter 3 of this thesis examines the factors influencing bat mortality at the St. Leon Wind Energy Project. Objectives: (1) to see if night-to-night variation in bat mortality is correlated with weather patterns in the vicinity of the turbines; (2) to see if night-to-night variation in mortality is correlated with insect abundance in the vicinity of the turbines; (3) to see if night-to-night patterns in mortality are correlated with bat activity at turbines; and, (4) to see if vegetation at the base of turbines significantly influences the accuracy of fatality estimates. Methods: Carcass surveys were conducted for 31 days in 2008 and 79 days in 2009. Turbines were selected randomly and carcass searches were conducted by a team of two searchers. Searchers surveyed a 40-m-radius circular plot at the base of each turbine. Searcher efficiency tests were conducted using bat carcasses at mowed and non-mowed turbines. Scavenging trials were conducted at mowed turbines in 2009. An effective daily-fatality rate was calculated for each turbine. Weather variables, insect abundance, and acoustic-activity data were also collected and used as predictor variables for mortality using generalized linear models.

Key Findings:

- In 2008, 55 bat carcasses were recovered and 32.7 percent were silver-haired bats, 52.7 percent were hoary bats, and 9.1 percent were eastern red bats.
- In 2008, adjusted mortality rate over the study period was 11.5 plus or minus 4.9 bats/turbine (7.2 plus or minus 3.1 bats/MW) for silver-haired bats, 15.7 plus or minus 5.5 bats/turbine (9.9 plus or minus 3.4 bats/MW) for hoary bats, and 2.0 plus or minus 1.0 bats/turbine (1.2 plus or minus 0.6 bats/MW) for eastern red bats.
- In 2009, only 33 bat carcasses were recovered and 57.6 percent were silver-haired bats, 36.4 percent were hoary bats, and 6.1 percent were eastern red bats.
- In 2009, adjusted mortality estimates were 7.7 plus or minus 2.1 bats/turbine (4.6 plus or minus 1.3 bats/MW) for silver-haired bats,

5.8 plus or minus 2.1 bats/turbine (3.5 plus or minus 1.3 bats/MW) for hoary bats, and 1.1 plus or minus 0.8 bats/turbine (0.7 plus or minus 0.5 bats/MW) for eastern red bats.

- No weather variables measured were correlated with bat mortality in 2008.
- In 2009, mortality of silver-haired bats was highest on calm, cold, humid nights when insect abundance was low.
- A significant positive correlation was found between bat activity measured at ground level and mortality of silver-haired and hoary bats in 2009.
- Adult males dominated mortality numbers in 2008; however, in 2009, most mortalities were subadult silver-haired bats and adult female hoary bats.
- Fatality estimates were considerably less reliable when vegetation was high.

Conclusions: Although the author was not able to directly test if tree bats were mating at communication towers or turbines, the results from this study were consistent with the Reproductive Landmarks Hypothesis. The author concluded that this study showed evidence that tree bats exhibit an attraction to tall structures during fall migration, the same time of year that they are killed at wind turbines. Mortality of silver-haired bats was high on calm, cold, humid nights when insect abundance was low.

Johnson, G.D., 2005, A review of bat mortality at wind-energy developments in the United States: Bat Research News, v. 46, p. 45–49.

This review describes the patterns of bat mortality at wind-energy developments in the United States as of 2004. In 2004, wind-power capacity in the United States was 6,740 MW, or enough to supply electricity to over 1.6 million homes.

- Eleven of the 45 species of bats in North America were included among the fatalities at wind farms, although none of these were classified as threatened or endangered by the federal government.
- Most (83.2 percent) fatalities involved migratory tree bats. Hoary bats (*Lasiurus cinereus*) comprised 45.5 percent of the 1,440 bats identified to species. Eastern red bats (*L. borealis*) were 26.3 percent of the fatalities and silver-haired bats (*Lasionycteris noctivagans*),

11.4 percent. The remaining fatalities were tri-colored bats (formerly eastern pipistrelles; *Perimyotis subflavus*, 8.5 percent), little brown bats (*Myotis lucifugus*, 5.9 percent), and big brown bats (*Eptesicus fuscus*, 1.9 percent). Six northern long-eared bats (*Myotis septentrionalis*) and single specimens of the western red bat (*Lasiurus blossevillii*), Brazilian free-tailed bat (*Tadarida brasiliensis*), long-eared myotis (*Myotis evotis*), and Seminole bat (*Lasiurus seminolus*) were also found killed at U.S. wind farms.

- In the Upper Midwest (UM), hoary and eastern red bats comprised most fatalities, whereas eastern red bat, tricolored bats, and hoary bats were most commonly killed in the East. In the Pacific Northwest (PNW), equal numbers of hoary and silver-haired bats were killed, whereas most fatalities in the Rocky Mountain West (RMW) were hoary bats.
- Approximately 90 percent of fatalities occurred from mid-July through late September. Collisions were lower during the breeding season and during spring migration.
- As of 2004, 11 studies had been conducted in the United States to estimate fatality rates of bats at wind farms. Estimated fatality rates adjusted for searcher efficiency and scavenger removal were 1.2 bats/turbine in the PNW and RMW, 1.7 bats/turbine in the UM, and 46.3 bats/turbine in the East.
- Studies suggested that mortality rates were highest in forested environments, moderate in open areas close to forests, and lowest in open areas.
- Inclement weather did not appear to correlate with mortality of bats.
- Warning lights for airplanes that were placed on turbines did not increase fatalities.

Johnson, G.D., Erickson, W.P., Strickland, M.D., Shepherd, M.F., and Sarappo, S.A., 2003, Mortality of bats at a large-scale wind power development at Buffalo Ridge, Minnesota: American Midland Naturalist, v. 150, p. 332–342.

This paper documents mortality of bats at a large-scale wind-power development at Buffalo Ridge, southwestern Minnesota, from 1996–1999. Objectives: (1) to estimate the number of bat mortalities attributable to collisions with wind turbines for the entire Buffalo Ridge wind-energy facility; (2) to determine the species and groups at highest risk; and, (3) to determine what factors may be related to collision mortality. There were three phases to this study and the phases were based on the installation of different types of turbines. Methods: Fatality searches were conducted at a select number of turbines during each phase of the project. Each turbine was searched every 14 days from March 15 to November 15 each study year. A 100-m×100-m square plot was centered around each turbine and variable transect widths were used based on visibility within each habitat type. Carcass-removal trials and searcher-efficiency trials were conducted to estimate the time that bat fatalities remained in the search area and to estimate a rate of searcher efficiency.

Key Findings:

- Of the 163 bat fatalities that could be identified to species, hoary bats (*Lasiurus cinereus*) comprised 66 percent, and eastern red bats (*L. borealis*) comprised 23 percent. The remaining fatalities were silver-haired bats (*Lasionycteris noctivagans*), tricolored bats (*Perimyotis subflavus*), little brown bats (*Myotis lucifugus*), and big brown bats (*Eptesicus fuscus*).
- Both the hoary and eastern red bat samples were primarily males.
- Ninety-seven percent of the fatalities occurred from July 15 to September 15.
- Injuries sustained by bats included fractured wings, legs and necks; head wounds; and abrasions and abdominal injuries.
- Bat fatalities were fairly widespread throughout the study area.
- Fifty-four percent of all bat carcasses were found less than or equal to 10 m from a turbine, 43 percent were found from 10 m to 20 m, 3 percent were found from 20 m to 30 m and one (0.5 percent) was found greater than 30 m from a turbine.
- The mean number of fatalities at lighted turbines was not significantly higher than the mean number of fatalities at unlit turbines.
- For all three phases of the study, 541 bat collision fatalities were estimated to occur each year.

Johnson, G.D., Perlik, M.K., Erickson, W.P., and Strickland, M.D., 2004, Bat activity, composition and collision mortality at a large wind facility in Minnesota: Wildlife Society Bulletin, v. 32, p. 1278– 1288.

The authors investigated bat activity, composition, and collision mortality at the Buffalo Ridge wind farm in southwestern Minnesota from June 15 to September 15, 2001 and 2002. Objectives: (1) to estimate relative abundance and composition of bat populations in the study area; (2) to estimate the number of bat fatalities attributable to wind-energy-facility development; (3) to determine whether the mortality was occurring to resident breeders, migrants, or both; and, (4) to determine what physical (for example, turbine characteristics) and biological (for example, habitat) factors might be associated with bat collision mortality in the study area. Methods: Bat activity was monitored using Anabat II bat detectors at potential foraging and roosting sites within commuting distance of the wind-energy facilities. To monitor bat activity within the wind-energy facilities, detectors were placed at each of three turbines each survey night and selection of turbine locations were made using a systematic design with a random start and removed to different turbines each night. Mist-netting surveys were conducted in woodlands and over ponds and other wetlands within 3.6 kilometers (km) of the wind facilities to obtain data on bat-species composition in the study area. Fatality searches were conducted by searching for carcasses daily at the three turbines where the Anabat detectors were placed the previous night. This activity was performed to determine relationships between bat activity at turbines and collision mortality levels. Eighty (in 2001) and 100 (in 2002) additional turbines were selected for fatality searches using a systematic design and random start. Turbines were searched every 14 days throughout the study. Fatality estimates were adjusted for removal bias and searcher efficiency. Poisson regression models were used to investigate the relationship between bat activity and mortality levels and cover type within 100 m of the turbines, distance from each turbine to the nearest wetland and woodland, and presence of FAA lighting.

Key Findings:

- More bat activity was found at foraging and roosting areas than at turbines.
- Peak bat activity at turbines followed the same trend as mortality and occurred from mid-July through the end of August.
- Distance to woodland was the only landscape variable significantly related to bat activity, and bat activity decreased with increasing distance from woodlands.
- There was no significant difference between number of bat passes and whether turbines were lighted.
- There was no relationship between bat activity and bat mortality at turbines.

- Mist-netting surveys captured big brown bats (64), silver-haired bats (17), little brown bats (11), eastern red bats (8), and hoary bats (3).
- Fatality searches found 151 bat casualties during the study, including 115 hoary bats, 21 eastern red bats, 8 big brown bats, 4 silver-haired bats, and 3 little brown bats.
- Eighty-two percent of fatalities occurred during the period July 16–August 31.
- Bat collisions occurred during inclement weather as well as during clear weather.
- Mean number of bat fatalities did not differ between lighted and unlighted turbines.
- Estimated total bat mortality for both wind facilities combined was 849 in 2001 and 364 in 2002, which equates to an estimated mean of 3.02 fatalities/turbine in 2001 and 1.30 fatalities/turbine in 2002.

Conclusions: Mortality of bats at wind turbines does not appear to involve foraging bats. The peak in collision mortality corresponds to the period of post-breeding southward migration for hoary, eastern red, and silver-haired bats. Both Anabat and mist-net data indicate that relatively large breeding populations of bats were close to the wind facility when collision mortality was low to nonexistent.

Johnson, J.B., Gates, J.E., and Zegre, N.P., 2011a, Monitoring seasonal bat activity on a coastal barrier island in Maryland, USA: Environmental Monitoring and Assessment, v. 173, p. 685–699.

This study used acoustic monitoring to examine the seasonal bat activity on the Assateague Island National Seashore, a barrier island off the coast of Maryland, from 2005 to 2006. Objectives: (1) to establish seasonal bat activity patterns on Assateague; (2) to determine if bats use the barrier island during migration; and, (3) to determine if bat-activity patterns were correlated with weather conditions. Methods: Anabat II detectors were used to monitor bat echolocation passes at three bat-monitoring stations. The total number of bat passes was summed for each monitoring station as an index of bat activity. A time-series analysis was used to account for weather and to model nightly and seasonal bat passes. The time-series analysis accounted for autocorrelation. Generalized least squares was used to develop a predictive model of bat activity and to account for seasonal and climatic influences.

Key Findings:

- 39,384 bat passes were recorded from five species, including eastern red bats (59.33 percent of total bat passes), big brown bats (3.05 percent), hoary bats (0.10 percent), tricolored bats (0.10 percent), and silver-haired bats (0.07 percent).
- Bat passes generally increased beginning in April, peaked in August, and declined gradually until December, when relatively few passes were recorded.
- After accounting for seasonal trends and autocorrelation, all weather variables explained some residual variation in bat activity. However, the degree to which weather variables accounted for bat activity varied among the monitoring stations in terms of significance. Mean nightly temperatures consistently had a significant positive effect on bat activity, and mean nightly wind speeds consistently had a negative effect on bat activity.

Conclusions: This research concluded that bats migrate along the Atlantic coastline and use barrier islands, possibly for navigation or as stopover sites. This has implications for proposed offshore wind-energy facilities, in that bats migrating along the coast may collide with wind turbines. Wind turbines offshore may be attractive to bats as resting places or possible mating areas; this would increase the risk of bat mortality. Bats were less active during periods of low temperatures and high wind speeds, possibly above thresholds for wind turbine cut-in speed.

Johnson, J.S., Watrous, K.S., Giumarro, G.J., Peterson, T.S., Boyden, S.A., and Lacki, M.J., 2011b, Seasonal and geographic trends in acoustic detection of tree-roosting bats: Acta Chiropterologica, v. 13, p. 157–168.

This study examined seasonal and geographic trends in acoustic detection of migrating tree-roosting bats at 1 existing and 13 proposed wind-energy facilities in seven eastern states in the United States between April and November of 2007 and 2008. Hypotheses: (1) nightly detection rates of migratory tree-roosting species would peak earlier in the northeastern United States during fall migration and peak earlier in the mid-Atlantic United States during spring migration; (2) nightly species-specific detection rates would be correlated between surveys located in the same geographic region; and, (3) nightly species-specific detection rates would be correlated with nightly species mortality at the nearby wind energy facility. Methods: In 2007, acoustic bat surveys were conducted in Maine (2), New Hampshire (1), Pennsylvania (3), and West Virginia (1). In 2008, surveys were conducted in Maine (2), Vermont (1), New York (1), Pennsylvania (3), West Virginia (2), and Virginia (1). Each survey was placed into one of three geographic regions to allow for latitudinal comparisons: northeast (Maine, New Hampshire, and Vermont), northern Allegheny Plateau (southern New York and northern Pennsylvania), and mid-Atlantic (southern Pennsylvania, West Virginia, and Virginia). Anabat II detectors were used to record echolocating bats. Species-specific linear correlations were used to compare nightly silver-haired, hoary, and eastern red bat calls/detector/night between all possible pairs of survey locations within the same geographic regions. Nightly silver-haired, hoary, and eastern red bat calls/detector/night at two proposed sites in West Virginia were compared to the number of mortalities at Mount Storm Wind Energy Facility, also in West Virginia, using linear correlations.

Key Findings:

- Nightly detection rates of silver-haired and hoary bats at the two sites in West Virginia were positively correlated with the number of dead silver-haired and hoary bats at the Mount Storm Wind Energy Facility.
- No significant correlation was found between the eastern-red-bat detection rate and the mortality at Mount Storm Wind Energy Facility.
- Nightly detection rates of silver-haired, hoary, and eastern red bats during fall migration peaked earlier in the year in the two northernmost geographic regions compared to the mid-Atlantic region.
- Nightly detection rates of the three species during the spring migration peaked earlier in the year in the mid-Atlantic compared to geographic regions farther north.
- In all regions, eastern red bats were more frequently detected later in the year than were hoary bats. Conclusion: The seasonal detection rates of hoary bats, silver-haired bats, and eastern red bats reflect their different migratory patterns, and these patterns may be useful in predicting the timing of mortality events at wind-energy facilities.

Klug, B.J., and Baerwald, E.F., 2010, Incidence and management of live and injured bats at wind-energy facilities: Journal of Wildlife Rehabilitation, v. 30, p. 11–16.

During studies of patterns and causes of bat fatalities at wind-energy facilities, the authors found numerous live and injured bats. Objectives: to examine the patterns of live-bat occurrences at three wind-energy facilities in southwestern Alberta, Canada, and to suggest methods for conducting searches at wind-energy facilities to increase the likelihood of finding live bats. Methods: Carcass searches were conducted at 40 turbines from July 15 through September 30, 2006, and at 42 turbines from July 15 through September 30, 2007. Two searchers used a spiral technique with a 45-m rope to search the base of turbines. If a live bat was found, it was immediately immobilized and transported to a vehicle for assessment and treatment. Bats were examined for obvious physical trauma such as hemorrhaging, broken or severed wings, and open lacerations. Bats were either euthanized if the injuries were deemed too severe or they were rehydrated and monitored for recovery if the injuries did not seem life-threatening.

Key Findings:

- Of 1,033 bats encountered during the two years of the study, 26 (2.5 percent) were alive.
- Significantly more live bats were found at turbines when searched daily than turbines searched weekly.
- Only eight of the live bats found had visible signs of skeletal damage or considerable soft tissue trauma.
- Injuries observed included wing fractures, wing amputation, leg fractures, and open abdominal wounds.
- Sixteen of the 26 live bats found were treated and 13 (81.3 percent) were successfully released. Conclusions: Searching turbines daily results in the discovery of more live bats than does searching each turbine once per week. Dehydration appeared to be the immediate threat to survival for bats without severe injuries. The authors suggest that live and injured bats can be properly managed if personnel involved in fatality surveys at wind-energy facilities are authorized to handle live bats and are prepared to rehydrate suitable individuals and euthanize those with terminal injuries.

Klug, B.J., Turmelle, A.S., Ellison, J.A., Baerwald, E.F, and Barclay, R.M.R., 2011, Rabies prevalence in migratory tree-bats in Alberta and the influence of roosting ecology and sampling method on reported prevalence of rabies in bats: Journal of Wildlife Diseases, v. 47, p. 64–77.

This study examined rabies prevalence in hoary (*Lasiurus cinereus*) and silver-haired bats (*Lasionycteris noctivagans*) collected during mortality surveys at wind-energy facilities in southern Alberta, Canada, during 2007 and 2008. Objective: to estimate rabies prevalence in hoary and silver-haired bats obtained from carcasses collected at wind-energy facilities and compare this data to prevalence estimates reported in the literature. Methods: Bat carcasses were collected during mortality surveys conducted at three wind-energy facilities in 2007 and 2008 (see Baerwald and Barclay, 2009). When a bat carcass was discovered, it was identified to species, sexed, and aged. Carcasses were submitted for rabies diagnosis to the Centers for Disease Control and Prevention (CDC) in Atlanta, Georgia. A literature review was conducted on both peer-reviewed and government reports for reported rabies prevalence in bats in North America. Rabies rates were compiled for passive surveillance from public health records spanning the previous 56 yr, and were also obtained for active-surveillance estimates. Estimates of rabies prevalence from this study was compared to rabies prevalence in the literature using Chi-square or Fisher's exact tests.

Key Findings:

- Rabies testing yielded a single positive result from a total of 121 hoary bats and 96 silver-haired bats submitted for testing. The positive bat was a subadult silver-haired bat and was typed as the silver-haired bat variant.
- For both species, rabies prevalence was lower than the combined estimates of prevalence reported from passive surveillance.
- The prevalence estimates in this study did not differ from previous estimates for these two species based on active-surveillance studies.
- Rabies-prevalence estimates based on passive surveillance were higher than estimates from active surveillance across all bat species.
- Non-synanthropic bat species had higher rabies prevalence than synanthropic species.

Conclusions: Most previously reported rabies-prevalence estimates for hoary and silver-haired bats are overestimates due to a bias in sampling. Furthermore, high rabies-prevalence estimates in some bat species are partly due to a bias in sampling resulting from the difference in roosting ecology. The authors also point out that the influx of bat carcasses from wind facilities could have effects on the surrounding ecosystem by potentially increasing the risk of disease transmission to mammalian scavengers.

Kunz, T.H., Arnett, E.B., Erickson, W.P., Hoar, A.R., Johnson, G.D., Larkin, R.P., Strickland, M.D., Thresher, R.W., and Tuttle, M.D., 2007a, Ecological impacts of wind energy development on bats—Questions, hypotheses, and research needs: Frontiers in Ecology and the Environment, v. 5, p. 315–324.

This paper synthesizes information on the impacts of wind-energy development on bats. The authors highlight ongoing development of wind-energy facilities in the United States, summarize evidence for bat fatalities at these sites, make projections of cumulative fatalities of bats for the Mid-Atlantic Highlands (Maryland, Pennsylvania, Virginia, and West Virginia), identify research needs to help reduce or mitigate adverse environmental impacts at these facilities, and propose hypotheses to evaluate where, when, how, and why bats are being killed.

Highlights:

- Since utility-scale wind turbines were first deployed in the United States in the 1980s, turbine height and the rotor-swept area has steadily increased with each new generation of turbines.
- Relatively small numbers of bat fatalities were reported at wind-energy facilities in the United States prior to 2001.
- Recent monitoring studies have indicated that some wind-energy facilities have killed large numbers of bats.
- Of the 45 species of bats found in North America, 11 have been identified during ground searches.
- Nearly 75 percent of carcasses found were foliage-roosting eastern red bats, hoary bats, and tree cavity-dwelling silver-haired bats, each of which migrate long distances.
- Number of bats killed in the eastern United States at wind-energy facilities installed along forested ridgetops has ranged from 15.3 to 41.1 bats/MW/yr. Bat fatalities reported from other regions of the western and mid-western United States have been lower, ranging from 0.8 to 8.6 bats/MW/yr.
- Most bat fatalities in North America have been reported in late summer and early autumn.
- Highest bat fatalities occur on nights when wind speed is low (less than 6 m s^{-1}).
- Bats are being struck and killed by the turning rotor blades of wind turbines; however, it is unclear why wind turbines are killing bats.

- Questions of why bats are killed at turbines include: Are bats attracted to wind turbines? Are bats attracted to sites that provide rich foraging habitats? Are bats attracted to the sounds produced by wind turbines? Do some weather conditions place bats at increased risk of being killed by turbines? Are bats at risk because they are unable to acoustically detect the moving rotor blades?
- The projected number of annual fatalities (rounded to the nearest 500) in the year 2020 range from 33,000 to 62,000 individuals, based on the National Renewable Energy Laboratory (NREL) WinDS model. Using the PJM (Pennsylvania, Jersey, Maryland Power Pool) grid operator interconnection queue, the estimates of annual fatalities range from 59,000 to 111,000 bats. This breaks down to 9,500 to 32,000 hoary bats; 11,500 to 38,000 eastern red bats; and 1,500 to 6,000 silver-haired bats.
- There may be a substantial impact on both migratory and local bat populations; however, baseline population estimates and demographics are unknown for migratory tree bats.
- The authors propose 12 research directions in this paper and highlight that research should focus on regions and at sites with the greatest potential for adverse effects.
- The authors call for full cooperation and research support from the wind industry.
- Scientifically sound research and monitoring studies are needed to inform policy makers during the siting, permitting, and operation of renewable energy sources.

Kunz, T.H., Arnett, E.B., Cooper, B.M., Erickson, W.P., Larkin, R.P., Mabee, T., Morrison, M.L., Strickland, M.D., and Szewczak, J.M., 2007b, Assessing impacts of wind-energy development on nocturnally active birds and bats—A guidance document: Journal of Wildlife Management, v. 71, p. 2449–2486.

This publication complements and extends the National Wind Coordinating Committee document "Studying Wind Energy/Bird Interactions: A Guidance Document" (Anderson and others, 1999). The purpose of this document is to provide researchers, consultants, decision-makers, and other stakeholders with guidance to methods and metrics for investigating nocturnally active birds and bats in relation to utility-scale wind-energy development. The paper is organized with the following topics: (1) Limits of current knowledge about impacts of nocturnally active birds and bats; (2) Methods and metrics for observing nocturnal behavior of birds and bats; (3) Methods and metrics for collecting additional data on nocturnally active birds and bats; (4) Conducting pre- and post-construction monitoring; and, (5) Management implications. The authors emphasize that sampling protocols and methodology employed to quantify bird and bat mortalities must be rigorous and scientifically valid. Overall, these guidelines were developed to encourage efficient, cost-effective experimental designs that will result in comparable data.

Kuvlevsky, W.P., Jr., Brennan, L.A., Morrison, M.L., Boydston, K.K., Ballard, B.M., and Bryant, F.C., 2007, Wind energy development and wildlife conservation: challenges and opportunities: Journal of Wildlife Management, v. 71, p. 2487–2498.

This review paper examines wind-energy development and wildlife-conservation issues focusing on the Lower Gulf Coast (LGC) of Texas, a region identified as having potentially negative impacts on migratory birds and bats, with respect to wind-farm development. The authors summarize what is known about how wind-energy developments influence wildlife populations and identify specific research opportunities and management challenges that can help mitigate the potential negative effects of wind-energy development. An overview of bird-collision mortality and habitat loss is presented. Bird collisions range from 0 to greater than 30.0 collisions/turbine/year. This variability is due to numerous factors, including: inconsistent experimental designs and data-collection protocols, layout design of the wind farm, weather conditions, topography, and bird species and the numbers of birds using the site and their behavior. Habitat loss associated with wind energy-developments can be an even greater threat to bird populations than collision fatalities. An overview of bats and the impact of wind-farm development on their populations is presented. Other wildlife have the potential to be impacted by wind-energy developments (including other birds indirectly affected, mammals, and herpetofauna), because developments alter wildlife habitat. These alterations include habitat disturbance, construction of additional electrical transmission lines, and the development of extensive road networks. The authors conclude by highlighting the LGC region in Texas and its unique vegetation communities and topography. Three migratory bird flyways converge north of Corpus Christi, Texas. This flyway funnels tens of millions of migratory birds along the LGC each year. Wind farms would severely affect these migrating birds. In addition to birds, wind farms located in LGC could negatively impact resident and migratory bat populations, notably the migratory Mexican free-tailed bat (*Tadarida brasiliensis*). The authors conclude with a list of research needs and a discussion of experimental designs for investigating wildlife and wind energy interactions. Several management recommendations are provided: (1) develop guidelines to assist wind developers in how best to site their projects; (2) provide a place to house the compiled data from all wind developments in each state; and, (3) provide incentives for the developers to avoid high-risk areas within each state, and if these high-risk areas cannot be avoided, provide ways to offset or mitigate the effects of developing in high-risk areas.

Larkin, R.P., 2006, Migrating bats interacting with wind turbines—What birds can tell us: Bat Research News, v. 47, v. 23–32.

Small, night-migrating birds have been used as surrogates for bats when formulating hypotheses and suggesting research techniques for investigating migration patterns, but the appropriateness of this surrogacy is unknown. This purpose of this paper is to provide bat biologists with information on aspects of bird migration that might be useful in designing research on bat migration and wind-energy interactions with bats. Techniques for studying migration include searching for killed animals beneath tall structures, analyzing DNA and other chemical clues, making observations at night with optical instruments and radar, and banding bats. The author proceeds to describe the process of bird migration as it might pertain to bat migration. Based on ornithological data, bats would not be expected to fly in narrow corridors over level terrain or large bodies of water. However, when encountering mountains, some nocturnal migrants alter their course to fly along valleys and ridges, where winds and topography interact. There is enormous temporal variability in migration patterns for birds, and this should also apply to bats; studies on migration should not be short-term. There is also variability in the heights at which birds migrate. The height of bats engaged in long-distance migratory flight is also critical to understanding risks to mortality from wind turbines. Migrating bats may select narrowly defined height bands for favorable winds, and turbine blades may cut the air progressively above, at, or below this band. The author concludes by stating that biologists have been studying bird migration for many years and much of what is known may be useful in understanding the process of migration in bats (however, see Willis and others, 2010).

Miller, A., 2008, Patterns of avian and bat mortality at a utility-scaled wind farm on the southern High Plains: Lubbock, Tex., Texas Tech University, Unpublished Master's Thesis. 113 pp.

This thesis examines the patterns of avian and bat mortality at a utility-scaled wind farm at the Red Canyon Wind Energy Center in the southern portion of the Texas Panhandle. Chapter 1 provides a general introduction to wind-energy development and its effects on birds and bats. Chapter 2 looks at avian mortality patterns. Chapter 3 examines bat mortality patterns. This summary focuses on Chapter 3. Objectives: (1) assess the incidence and frequency of bat mortality; (2) model mortality rates; and, (3) identify the spatial and temporal distribution of bat mortality within the wind-energy development. Methods: Standardized carcass searches were conducted for 52 weeks, beginning in September 2006, on 28 of the 56 turbines. Bias corrections were made by conducting searcher efficiency and carcass-removal trials. Bias-correction trials were conducted seasonally to account for differences in vegetation and scavenger-predator characteristics. Spatial and temporal analyses were conducted using Poisson models.

Key Findings:

- During the study period, 56 bat mortalities were found.
- Brazilian free-tailed bats composed 94 percent of fatalities, hoary bats 4 percent, and red bats 2 percent.
- Mortalities were found beneath 24 of the 28 searched turbines.
- Peak mortality detections occurred in October.
- For Brazilian free-tailed bats, 41 percent of fatalities were adults, 43 percent were juveniles, and 16 percent unidentifiable to age. Fifty-three percent were male, 10 percent were female, and 37 percent were unidentifiable to sex.
- For hoary bats, 40 percent were adults, 40 percent were juveniles, and 20 percent were unidentifiable to age. Sixty percent were male, 20 percent female, and 20 percent were unidentifiable to sex.
- Turbines were classified as low, medium, and high mortality. Low-mortality turbines were estimated to kill 15.29 bats/MW. Medium-mortality turbines were estimated to kill 45 bats/MW, and high-mortality turbines were estimated to kill 106.67 bats/MW.
- Total mortality was estimated for the entire wind energy development as 36.87 bats/MW. Conclusions: Bat mortality at this site was above the reported average bat mortality rate reported for new-generation

wind projects in the United States (2.1 mortalities per MW per year). This is one of the first post-construction mortality monitoring studies conducted on wind energy development in Texas and the southwestern United States.

Osborn, R.G., Higgins, K.F., Dieter, C.D., and Usgaard, R.E., 1996, Bat collisions with wind turbines in southwestern Minnesota: Bat Research news, v. 37, p. 105-108.

Prior to this study, most studies on the biological effects of wind energy were designed to assess effects on birds. Bat collisions with turbines appeared to be infrequent and had therefore received little attention by natural-resource agencies or the scientific community. While conducting a study to assess the impacts of wind energy on birds in southwestern Minnesota, the authors found several bats. Objectives: (1) to document the species composition and timing of bat mortality associated with the Minnesota wind facility during 1994 and 1995; and, (2) to review some current theories regarding bat collisions with anthropogenic structures and to speculate on the possible causes of bat mortality at wind turbines. Methods: Buffalo Ridge Wind Resource Area (WRA) was searched for wildlife carcasses from April 1994 through December 1995. Search intensity varied by season: during spring and fall, all 73 turbine plots were searched once per week; however, during summer and winter, only 30 plots were searched once per week.

Key Findings:

- During 20 months of continuous monitoring, 13 bat carcasses were found, representing 5 species.
- Five of the carcasses found were hoary bats, four were silver-haired bats, two were eastern red bats, and one each of a big brown bat and a tricolored bat.
- Carcass-recovery dates ranged from May 12 to September 22. One collision (8 percent of total) occurred in spring, 11 (85 percent) during the summer, and one (8 percent) on the first day of fall.
- Six of the 13 bat carcasses (46 percent) were found within 15 m of turbines, and 69 percent were found within 20 m.
- On September 13, 1995, a tricolored bat carcass was found, extending the known range of this bat approximately 160 km farther west in Minnesota.

Conclusions: The timing of bat-carcass recovery at Buffalo Ridge WRA suggests that bats are susceptible to collisions with anthropogenic structures at times other than during migration. The findings from this study represent the first evidence of frequent bat collisions with wind turbines.

Piorkowski, M.D., and O'Connell, T.J., 2010, Spatial pattern of summer bat mortality from collisions with wind turbines in mixed-grass prairie: American Midland Naturalist, v. 164, p. 260–269.

This study examined the spatial pattern of summer bat mortality from collisions with wind turbines in the southern Great Plains, Oklahoma, during May–July, 2004 and 2005. The authors proposed to test the hypothesis that collision mortality was randomly distributed among the 68 wind turbines at the Oklahoma Wind Energy Center (OWEC). Objectives: (1) to estimate a summer-season, per-turbine rate of bat collision mortality; and, (2) to analyze the spatial distribution of collisions at the OWEC relative to vegetation cover, topography, and location of individual turbines. Methods: The bases of all 68 turbines were visually inspected for bird and bat carcasses in 2004 and 2005. Surveys occurred in late May and late June in 2004 and every 2 weeks from May 15–July 15 in 2005. Circular transects were walked around the base of each turbine at 5, 10, and 15 m from the base. An observer-detection rate was estimated in 2005 with blind field trials and models built to approximate free-tailed bats. Three estimates of mortality were calculated based on (1) raw carcass counts, (2) raw counts divided by observer-detection rate, and (3) raw counts modified by observer-detection rate and removal rate. Broad spatial distribution of mortalities were plotted using GIS, and "hotspot" analyses were conducted for fine-scale spatial distribution patterns.

Key Findings:

- Eleven turbine-killed birds of six species and 111 turbine-killed bats of seven species were found.
- First documented cave myotis (*Myotis velifer*) killed by a wind turbine.
- The majority of bat carcasses (94) were Brazilian free-tailed bats (*Tadarida brasiliensis*) representing 78 percent of all fatalities and 85 percent of all bat fatalities.
- Examination of 64 (58 percent) relatively fresh and intact bat carcasses indicated a female-biased sex ratio.

- Raw carcass counts yielded estimates of 0.53/fatalities/MW in 2004 and 0.56 fatalities/MW in 2005.
- For Brazilian free-tailed bats, rates adjusted for probability of detection and incorporating published removal estimates yielded 74–92 individuals (0.73–0.90/MW) in 2004 and 67–85 (0.66–0.83/MW) in 2005.
- Carcasses were detected up to 46 m from turbine bases; however, 77 percent were located within a 20-m search radius.
- Brazilian free-tailed bat mortality was not associated with topography or habitat when both years of the study were combined.
- Three turbines emerged as significant "hot spots" of collision mortality. These three turbines were located near the head of a forested ravine cut by an intermittent tributary to the Cimarron River. The three turbines were less than 5 percent of the turbines at OWEC, but they accounted for 16 percent of the Brazilian free-tailed bat carcasses collected during both years.
- Seventeen individuals were necropsied; 14 specimens had broken bones and 3 had no apparent skeletal injury. In the bats with no sign of injury, no obvious visceral hemorrhaging was observed potentially consistent with barotrauma.

Conclusions: The results from this study provide some of the first evidence for a steady rate of collision mortality of Brazilian free-tailed bats at a North American wind farm, most likely due to the site's proximity (roughly 15 km) to both a maternity colony and a bachelor colony. Authors rejected their original hypothesis of random collision mortality among the 68 turbines at OWEC. The cluster of three turbines emerged as a hotspot of collision mortality. This suggests that bats moving through this area were either more susceptible to collision or that the number of bats moving through the area was greater than at other areas of OWEC. They concluded that this provided support for the "coincidental collision" hypothesis of Cryan and Barclay (2009). If monitoring pointed to particular turbines as hotspots for mortality, then mitigating management (for example, curtailment of those specific units under specific conditions) could be developed.

Reynolds, D.S., 2006, Monitoring the potential impact of a wind development site on bats in the northeast: Journal of Wildlife Management, v. 70, p. 1219–1227.

This paper investigated spatial and temporal patterns of bat activity across a proposed wind-energy project site during the summer breeding season and the migratory season in the Tug Hill Plateau region of western New York. Hypotheses: (1) the physiogeography of the site would limit the species diversity and total bat abundance at the project site; (2) the bat community would be shifted towards species that are more commonly found at higher elevation (for example, hoary and silver-haired bats); and, (3) the climate of the project site would shift the sex ratio of the bat community towards males that were not as energy limited as reproductive females during the summer months. Methods: The study area was the Maple Ridge Wind Project, a 198-turbine project that began construction in August, 2005. Summer activity was sampled with mist netting and acoustical monitoring (Anabat II) in 2004. Migratory activity was sampled using acoustic monitoring during spring 2005, at two locations. Meteorological data was collected by the Maple Ridge wind-development group. Summer activity was analyzed using a 2-factor general linear model. Temporal patterns of bat activity during the spring migratory sampling period using a 3-factor general linear model.

Key Findings:

- Mist-netting was conducted at 24 sites and 35 bats of 3 species were captured. Northern long-eared bat (*Myotis septentrionalis*, 17), little brown bat (*M. lucifugus*, 15), and big brown bat (*Eptesicus fuscus*, 3).
- Summer acoustic sampling found that activity levels were highly skewed across sample sites.
- Bat activity was significantly influenced by habitat, with ponds being the only habitat showing preferential use by bats.
- Bats of the genus *Myotis* accounted for almost 95.7 percent of all bat calls and 98.8 percent of all feeding buzzes.
- Acoustic monitoring during the spring migratory period revealed a median activity level of 2.0 bats/night.
- On 20 April, 101 bat passes of the tricolored bat (*Perimyotis subflavus*) were recorded and on the 10 June, 115 bat passes of the hoary bat (*Lasiurus cinereus*) were recorded.
- Most of the migratory activity occurred at minimum wind speeds below 1.3 m/s.
- Temperature appeared to have a strong influence on migratory bat activity. Days with high migratory bat activity had a mean maximum temperature of 23.9 plus or minus 4.4°C compared to 9.8 plus or minus 4.8°C.

Conclusions: The results of this study strongly suggest that the Tug Hill Plateau region does not support a large bat population in terms of either species diversity or total bat abundance. Most of the bat migratory activity occurred when daily mean wind speeds were below 5.4 m/s. This is encouragingly close to the cut-in speed (the lowest economically useable wind speed) of a typical commercial wind turbine. (This study did not sample fall migration.)

Rollins, K.E., Meyerholz, D.K., Johnson, G.D., Capparella, A.P., and Loew, S.S., 2012, A forensic investigation into the etiology of bat mortality at a wind farm—Barotrauma or traumatic injury?: Veterinary Pathology, accessed February 14, 2012, *http://vet.sagepub.com/content/2012/01/30/0300985812436745*. [published online January 30, 2012]

The authors in this paper studied the causes of bat mortality at a central Illinois wind-energy facility compared to bat mortality from collisions with buildings in Chicago, Illinois (Ill.). They designed a 2-part forensic investigation to examine two competing hypotheses for the mortality of bats at turbines: (1) blunt force trauma from being struck by moving turbine blades and (2) barotrauma. Objectives: (1) to test whether postmortem decomposition and environmental conditions would influence the development of morphologic artifacts of pulmonary barotrauma using laboratory mice; and, (2) to examine salvaged bats collected from a wind-energy facility compared with a control group of bats that had died following collisions with buildings. Methods: The authors conducted two experiments. The first experiment used laboratory mice and performed a decomposition trial, an elevated temperature trial, and a freezing trial. After each trial, lungs, heart, and trachea were removed from each mouse and fixed for routine processing and staining. Lungs were examined for signs of vascular congestion, erythrocyte extravasation, and edema, and for each parameter, samples were scored for extent of damage. The second experiment examined 53 bats salvaged near buildings in downtown Chicago, Ill., and compared these with 262 bats salvaged from the wind-energy facility. Both bats killed by buildings and bats killed at the wind-energy facility were radio-graphically examined for bone lesions consistent with traumatic injury. Additionally, 66 of the 262 bats from the wind-energy facility were necropsied and examined for supplemental lesions consistent with trauma. All bats killed by buildings were assessed for external lacerations, but were not available for full necropsies. A subset of each group of bats was also examined for evidence of auditory barotrauma.

Key Findings:

- Decomposition caused changes in vascular congestion, erythrocyte extravasation, and edema in murine lungs. Vascular congestion scores increased 2 hr postmortem, edema scores increased after 24 hr, and erythrocyte extravasation scores increased after 72 hr.
- Elevated temperatures (33°C) accelerated postmortem morphologic changes in mice.
- Freezing and freezing/thawing trials increased erythrocyte extravasation and edema compared to fresh lung tissue.
- Bone fractures were detected in 9 of 53 (17 percent) building-killed bats, whereas 152 of 262 (58 percent) turbine-killed bats had bone fractures.
- Among bats without fractures, turbine-killed bats had significantly more cases with external lacerations than building-killed bats.
- Only one (1 of 42; 2 percent) building-killed bat showed evidence of tympana (ear drum) rupture, whereas 11 of 81 (14 percent) of turbine-killed bats showed evidence of ruptured tympana.

However, only 5 of 81 (6 percent) of the bats with tympana ruptures had no detectable evidence of traumatic injury; this suggests that a small fraction of turbine-killed bats had lesions possibly consistent with barotrauma.

Conclusions: Determining the causes of lung damage from salvaged bats can be confounded by common postmortem artifacts (for example, decomposition, temperature differences) as well as the competing hypothesized causes of traumatic injury (for example, blunt force trauma, barotrauma, laceration). Because these competing factors cannot be confidently excluded during a pathological examination, pulmonary barotrauma should not be diagnosed from salvaged bat carcasses unless assessments are also made for ruptured ear drums. The authors conclude that their results point to a need for improvements in the current monitoring procedures at wind-energy facilities and in the experimental approaches used to determine whether barotrauma or other causes contribute to bat mortality.

Smallwood, K.S., and Karas, B., 2009, Avian and bat fatality rates at old-generation and repowered wind turbines in California: Journal of Wildlife Management, v. 73, p. 1062–1071.

This study examined avian- and bat-fatality rates at old-generation and repowered wind turbines at Altamont Pass Wind Resource Area (APWRA) in California. The APWRA began operations during the 1980s and until recently was the world's largest wind farm. In 1998, the APWRA comprised about 5,400 wind turbines of various models. In 2005, 126 turbines were replaced with 31 horizontal-axis wind turbines. The new turbines were more widely spaced and operated at a lower rotor speed, which were traits thought by some to be safer for birds. Objectives: to compare estimates of fatality rates between (1) the periods 1998–2003 and 2005–2007; and, (2) a repowered wind project and the concurrently operating old-generation wind turbines. Methods: Fatality searches were performed at selections of turbines. Searches were performed by walking parallel transects about 4–8 m apart. Fatality rates were adjusted for carcasses not found due to searcher-detection error and for scavenger removal. Bat–fatality rates were estimated by applying scavenger-removal and searcher-detection rates estimated for small-bodied bird species. Therefore, estimates of bat–fatality rates were likely biased low.

Key Findings:

- For APWRA-wide fatality rates, between 1998–2003 and 20052007, estimated mean adjusted fatality rate increased 23 percent for all birds combined.
- Comparing only adjusted fatality rates from old-generation turbines mutually monitored both 1998– 2003 and 2005–2007, fatality rates increased 51 percent for all birds combined.
- The first repowering project in the APWRA did not change fatality rates for any species or group of specie, because fatality rates did not differ between the old vertical-axis turbines and the new horizontal-axis turbines.
- The adjusted fatality rate of bats increased from zero at the old vertical-axis turbines to 16.4/yr at the new, repowered turbines, but this difference was not significant, probably due to small sample sizes.
- Compared to the concurrently operating old-generation turbines during 2005–2007, adjusted fatality rates in the repowered turbines were 66 percent lower for all birds combined.
- The adjusted fatality rate of bats was nearly 800 percent greater at repowered turbines compared to the concurrently operated old-generation turbines, but this large difference was not significant, probably due to sample sizes.

Conclusions: Repowering the entire APWRA would likely reduce fatality rates a great deal for birds, especially if considered on a power-generation basis and if carefully done by locating new turbines where they pose the least hazard. A possible downside to repowering, however, may be increased bat fatalities caused by wind turbines. Extrapolating the mean adjusted bat-fatality rate from the repowered turbines to a completely repowered APWRA, about 454 bat fatalities/year might result, but using more realistic scavenger-removal and searcher-detection rates could increase this number to the thousands of bats. More research is needed to address bat fatalities at the APWRA.

Sovacool, B.K., 2009, Contextualizing avian mortality—A preliminary appraisal of bird and bat fatalities from wind, fossil-fuel, and nuclear electricity: Energy Policy, v. 37, p. 2241–2248.

This article examines the threats that wind farms pose to birds (and bats) when compared to threats from fossil-fuel and nuclear electricity. The authors conducted a survey of 600 studies, articles, and reports investigating avian deaths and wind farms published from 1998 to 2008. The authors address three common problems with these studies: (1) they rarely compare their results with studies of other wind farms to contextualize their estimates; (2) most do not compare the possible avian deaths from wind electricity with other sources, and when they do, studies typically do not compare them to other energy sources; and, (3) no studies attempted to calculate the number of avian deaths per kilowatt-hour from energy sources, so that more meaningful comparisons might be made between different forms of electricity supply. To address these shortcomings, the author assessed and compared the avian deaths per GWh from three electricity systems: wind farms, fossil-fueled power plants (coal, natural gas, and oil generators), and nuclear plants. Averaged over six wind farms, 339 turbines were responsible for 0.279 avian deaths per GWh. The assessment of avian mortality for fossil-fuel power plants was estimated to be 5.18 fatalities per GWh. Finally, the total avian deaths per GWh for nuclear power plants was estimated to be about 0.416. The author concludes that for wind turbines, the risk appears to be greater to birds striking towers or turbine blades and for bats suffering barotrauma. For fossil-fueled power stations, the most significant fatalities come from climate change, which is altering weather patterns and destroying habitats that birds depend on. For nuclear power plants, the risk is almost equally spread across hazardous pollution at uranium mine sites and collisions with draft cooling structures. Fossil-fueled facilities are about 17 times more dangerous to birds on a per GWh basis than wind and nuclear power stations, according to the author. Putting this in absolute terms, wind turbines may have killed about 7,000 birds in 2006 but fossil-fueled stations killed 14.5 million and nuclear power plants 327,000. Far more detailed, rigorous, and sophisticated analyses are called for in the future. This preliminary study has quite a few shortcomings that are further discussed by the author. [The author lumped birds and bats together for this analysis. Also note that Willis and others (2010) disputes the methods and findings of this paper.]

Weller, T.J., and Baldwin, J.A., 2011, Using echolocation monitoring to model bat occupancy and inform mitigations at wind-energy facilities: Journal of Wildlife Management, v. 9999, p. 1–13, accessed January 23, 2012, at *http://www.fs.fed.us/psw/publications/weller/psw_2011_weller001.pdf*.

This research article combines results of continuous echolocation and meteorological monitoring at multiple stations to model the conditions that explain the presence of migratory bats at a wind-energy facility in southern California. Objectives: (1) to characterize year-round patterns of bat activity at a wind-energy facility in southern California; (2) to determine survey effort necessary to characterize bat-activity levels at the facility; (3) to model bat presence on-site with respect to date and meteorological conditions; and, (4) to determine whether incorporation of a full suite of meteorological conditions could improve predictions of bat presence. Methods: This study took place at the Dillon Wind Energy Facility (DWEF) near Palm Springs, California. This site is within the San Gorgonio Pass Wind Resource Area, one of the oldest and largest wind-energy developments in the world. Bat activity was monitored using Anabat II detectors attached to four meteorological towers at three different heights (2 m, 22 m, and 52 m above the ground). Detectors were operated nightly for 17 months from October 25, 2007 through March 31, 2009. Bat passes were classified as either "LowF" or "HiF." Most migratory species of bats use LowF echolocation calls. Site-occupancy models were used to estimate the probability of LowF bat presence on a given night. Presence and detection probability was estimated as a function of detector height, mean nightly wind speed, wind direction, temperature, and proportion of moon illumination.

Key Findings:

- Overall LowF bat-passes rates increased with detector height from 2 m [0.14 bat passes/detector night (dn)] to 22 m (0.26 bat passes/dn), and 52 m (0.43 bat passes/dn).
- Passes of LowF bats comprised 49.2 percent, 83.6 percent, and 98.5 percent of bat passes recorded at 2 m, 22 m, and 98.5 m, respectively.
- Detection probabilities ranged from 0.027 (SE = 0.014) at 2 m to 0.564 (SE = 0.032) at 52 m.
- Mean wind speed had the highest relative importance for predicting LowF bat activity in all but 2 time periods.
- Mean nightly wind speeds were lower on nights when LowF bat passes were recorded than on nights when they were not in all time periods.
- Proportion of moon illumination was important for predicting bat activity in 4 time periods. Conclusions: Despite low overall bat-activity levels, the authors were able to build models that successfully predicted conditions when LowF bats were present. The models developed confirmed results from wind-energy developments in other regions and habitats which have demonstrated that bat activity is generally higher at lower wind speeds and higher temperatures. The results also highlight the importance of modeling conditions that explain bat activity on a seasonal basis. There were marked differences in activity levels seasonally, and the relative importance of explanatory variables also differed seasonally. The authors conclude that using this model-based approach could improve the effectiveness and efficiency of mitigations for bat fatalities at wind-energy facilities. If properly conducted, echolocation monitoring may provide a reliable index to expected bat fatality levels at wind-energy facilities.

Willis, C.K.R., Barclay, R.M.R., Boyles, J.G., Brigham, R.M., Brack, V.B., Jr., Waldien, D.L., and Reichard, J., 2010, Bats are not birds and other problems with Sovacool's (2009) analysis of animal fatalities due to electricity generation: Energy Policy, v. 38, p. 2067–2069.

This forum paper is a response to Sovacool's (2009) comparison of North American bird (avian) and, presumably bat (chiropteran) mortality resulting from three methods of electricity generation. Although bats figure prominently in the title of Sovacool's paper, the text is virtually devoid of data on bats. The authors of this response point out that bats are not birds. Bats occupy an entirely different ecological niche than almost all birds, since bats are the primary consumer of night-flying insects. It has become increasingly clear that, in general, mortality at wind turbines is an issue that affects bats more than birds, and it is bats that face the most widespread and worrisome species-level conservation consequence from wind turbines. Bats investigate wind turbines; birds do not. Migrating bats appear to be attracted to wind turbines, and there is little evidence that birds are attracted. Bats are also prone to depressurization injuries (barotraumas) at turbines, to which birds are less susceptible. Sovacool's (2009) estimate of the average number of birds killed per GWh of wind power is incorrect and omits a large body of easily accessible, published data. Sovacool also states that death rates of all flying animals from wind turbines have decreased in recent years; however, this is not true for bats. There is clear evidence that, as turbines have become taller, bat mortality has increased. Sovacool also implies that pre-construction monitoring and mitigation at proposed wind-energy sites reduce the risk of mortality for birds and bats, which is simply not true for bats. Sovacool reviewed 600 sources, but failed to reference the primary literature on bat fatalities at wind turbines. The authors of this forum paper conclude by pointing out that Sovacool advocates averaging data across many species, large areas, and long time periods to estimate the total number of animals killed by three different electricity systems. This approach fails to address the importance of spatial and temporal variation when contextualizing the conservation implications of different electricity systems. Sovacool's conclusions are based on fundamental errors and could influence energy policy in ways that are counter-productive for the wind energy industry and the conservation of both groups of flying vertebrates: birds as well as bats.

III. Additional References

Arnett, E.B., Huso, M.M.P., and Morrison, M.L., 2007, Renewable energy resources and wildlife impacts and opportunities: Transactions of the Seventy-second North American Wildlife and Natural Resources Conference, 72, Portl9and, Oreg., March 20–24, 2007, Proceedings: Gardners, Penn., Wildlife Management Institute, p. 65-95.

Avery, M., and Clement, T., 1972, Bird mortality at four towers in eastern North Dakota—Fall 1972: Prairie Naturalist, v. 4, p. 87–95.

Bainbridge, T.R., 1985, The committee on standards—Precision and bias: American Society for Testing and Materials Standardization News, v. 13, p. 44–46.

Barclay, R.M.R., 1993, The biology of prairie bats, *in* Proceedings of the Third Prairie Conservation and Endangered Species Workshop, Holroyd, G.L, Dickson, H.L., Regnier, M., and Smith, H.C., eds., Natural History Occasional Paper, no. 19: Provincial Museum of Alberta, Edmonton, Alberta, Canada, p. 353–357.

Barclay, R.M.R., and Harder, L.M., 2003, Life histories of bats—Life in the slow lane, *in* Kunz, T.H., and Fenton, M.B., eds.: Bat Ecology, Chicago, Ill., University of Chicago Press, p. 209–253. Boyles, J.G., Cryan, P.M., McCracken, G.F., and Kunz, T.H., 2011, Economic importance of bats in agriculture: Science, v. 332, p. 41–42.

Crawford, R.L., and Baker, W.W., 1981, Bats killed at a north Florida television tower—A 25-year record: Journal of Mammalogy, v. 62, p. 651–652.

Cryan, P.M., 2003, Seasonal distribution of migratory tree bats (*Lasiurus* and *Lasionycteris*) in North America: Journal of Mammalogy, v. 84, p. 579–593.

DeBlase, A.F., and Cope, J.B., 1967, An Indiana bat impaled on barbed wire: American Midland Naturalist, v. 77, p. 238.

Denys, G.A., 1972, Hoary bat impaled on barbed wire: Jack-Pine Warbler, v. 50, p. 63.

Ellison, L.E., O'Shea, T.J., Bogan, M.A., Everette, A.L., and Schneider, D.M., 2003, Existing data on colonies of bats in the United States—Summary and analysis of the U.S. Geological Survey's Bat Population Database, *in* O'Shea, T.J., and Bogan, M.A., eds., Monitoring trends in bat populations of the United States and Territories—Problems and prospects: U.S. Geological Survey Information and Technology Report 2003-0003, p. 127–237.

Erickson, J.L., and West, S.D., 2002, The influence of regional climate and nightly weather conditions on activity patterns of insectivorous bats: Acta Chiropterologica, v. 4, p. 17–24.

Fenton, M.B., 2001, Bats, revised edition: New York, New York, Checkmark Books.

Flemming, T.H., and Eby, P., 2003, Ecology of bat migration, *in* Kunz, T.H., and Fenton, M.B., eds., Bat Ecology: Chicago, Ill., University of Chicago Press, p. 156–208.

Ganier, A.F., 1962, Bird casualties at a Nashville TV tower: Migrant, v. 33, p. 58–60.

Gollop, M.A., 1965, Bird migration collision casualties at Saskatoon: Blue Jay, v. 23, p. 15–17. Grindal, S.D., and Brigham, R.M., 1998, Short-term effects of small-scale habitat disturbance on activity of insectivorous bats: Journal of Wildlife Management, v. 62, p. 996–1003.

Hall, L.S., and Richards, G.C., 1972, Notes on *Tadarida australis* (Chiroptera: Molossidae): Australian Mammalogy, v. 1, p. 46.

Hibbard, E.A., 1963, Another hoary bat found hanging on a fence: Journal of Mammalogy, v. 44, p. 265.

Hitchcock, H.B., 1965, Twenty-three years of bat banding in Ontario and Quebec: Canadian Field-Naturalist, v. 79, p. 4–14.

Howell, J.A., 1997, Bird mortality at rotor swept area equivalents, Altamont Pass and Montezuma Hills, California: Transactions of the Western Section of the Wildlife Society, v. 33, p. 24–29.

Iwen, F.A., 1958, Hoary bat the victim of a barbed wire fence: Journal of Mammalogy, v. 39, p. 438. Jain, A.A., 2005, Bird and bat behavior and mortality at a northern Iowa wind farm: Ames, Iowa, Iowa State University, Unpublished Master's Thesis. 107 pp.

James, R.D., 2003, Bird observations at the pickering wind turbine: Ontario Birds, v. 21, p. 84–97. James, R.D., and Coady, G., 2004, Bird monitoring at Toronto's Exhibition Place Wind Turbine: Ontario Birds, v. 22, p. 79–88.

Johnson, P.B., 1933, Accidents in bats: Journal of Mammalogy, v. 14, p. 156–157.

Johnson, G.D., Erickson, W.P., Strickland, M.D., Shepherd, M.F., Shepherd, D.A., and Sarappo, A., 2002, Collision mortality of local and migrant birds at a large-scale wind-power development on Buffalo Ridge, Minnesota: Wildlife Society Bulletin, v. 30, p. 879–887.

Kunz, T.H., Gauthreaux, S.A., Jr., Hristov, N.I., Horn, J.W., Jones, G., Kalko, E.K.V., Larkin, R.P., McCracken, G.F., Swartz, S.M., Srygley, R.B., Dudley, R., Westbrook, J.K., and Wikelski, M., 2008, Aeroecology—Probing and modeling the aerosphere: Integrative and Comparative Biology, v. 48, p. 1–11.

Kunz, T.H., Braun de Torrez, E., Bauer, D., Lobova, T., and Fleming, T.H., 2011, Ecosystem services provided by bats: New York Academy of Sciences, v. 1223, p. 1–38.

Mabee, T.J., Cooper, B.A., Plissner, J.H., and Young, D.P., 2006, Nocturnal bird migration over an Appalachian Ridge at a proposed wind power project: Wildlife Society Bulletin, v. 34, p. 682–690. Mumford, R.E., and Whitaker, J.O., Jr., 1982, Mammals of Indiana: Indiana University Press, Bloomington, Ind. 537 pp.

Peurach, S.C., and Dove, C.J., 2009, A decade of U.S. Air Force bat strikes: Human-Wildlife Conflicts, v. 3, p. 199–207.

Osborn, R.G., Dieter, C.D., Higgins, K.F., and Usgaard, R.E., 1998, Bird flight characteristics near wind turbines in Minnesota: American Midland Naturalist, v. 139, p. 29–38.

O'Shea, T.J., and Bogan, M.A., 2003, Monitoring trends in bat populations of the United States and territories—Problems and prospects, U.S. Geological Survey Information and Technology Report 2003–0003.

O'Shea, T.J., Bogan, M.A., and Ellison, L.E., 2003, Monitoring trends in bat populations of the United States and territories—Status of the science and recommendations for the future: Wildlife Society Bulletin, v. 31, p. 16–29.

Piorkowski, M.D., 2006, Breeding bird habitat use and turbine collisions of birds and bats located at a wind farm in Oklahoma mixed-grass prairie: Stillwater, Okla., Oklahoma State University, Unpublished Master's Thesis. 100 pp.

Sauders, W.E., 1930, Bats in migration: Journal of Mammalogy, v. 11, p. 225.

Smallwood, K.S., 2007, Estimating wind turbine-caused bird mortality: Journal of Wildlife Management, v. 71, p. 2781–2791.

Taylor, W.K., and Anderson, B.H., 1973, Nocturnal migrants killed at a central Florida TV tower— autumns 1969–1971: Wilson Bulletin, v. 85, p. 42–51.

Terres, J.K., 1956, Migration records of the red bat, *Lasiurus borealis*: Journal of Mammalogy, v. 37, p. 442.

Timm, R.M., 1989, Migration and molt patterns of red bats: Bulletin of the Chicago Academy of Sciences, v. 14, p. 1-7.

U.S. Energy Information Administration, 2012, Electric power monthly March 2012: Washington D.C., U.S. Energy Information Administration Report, accessed on April 3, 2012, at *http://www.eia.gov/cneaf/electricity/epm/epm_sum.html.*

Valdez, E.W., and Cryan, P.M., 2009, Food habits of the hoary bat (*Lasiurus cinereus*) during spring migration through New Mexico: Southwestern Naturalist, v. 54, p. 195–200.

Van Gelder, R.G., 1956, Echo-location failure in migratory bats: Transactions of the Kansas Academy of Science, v. 59, p. 220–222.

West, M.J., 1999, Stereological methods for estimating the total number of neurons and synapses— Issues of precision and bias: Trends in Neuroscience, v. 22, p. 51–61.

Wisely, A.N., 1978, Bat dies on barbed wire fence: Blue Jay, v. 36, p. 53.

Zinn, T.L., and Baker, W.W., 1979, Seasonal migration of the hoary bat, *Lasiurus cinereus*, through Florida: Journal of Mammalogy, v. 60, p. 634–635.

International References

Ahlen, I., 2002, Fladdermoss och faglar dodade av windkraftverk [Wind turbines and bats—A pilot study]: Fauna and Flora, v. 97, p. 14–22.

Ahlen, I., Baagoe, H.J., and Bach, L., 2009, Behavior of Scandinavian bats during migration and foraging at sea: Journal of Mammalogy, v. 90, p. 1318–1323.

Alcalde, J.T., 2003, Impacto de los parques eolicos sobre las poblaciones de murcielagos: Barbastella, v. 2, p. 3–6.

Bach, L., 2001, Fledermause und windenergienutzung—Reale probleme oder einbildung? [Bats and wind turbines—Real problems or only fancies?]: Vogelkdliche Berrichte aus Niedersachsen, v. 33, p. 119–124.

Bach, L., and Rahmel, U., 2004, Summary of wind turbine impacts on bats—Assessment of a conflict: Bremer Beitrage fur Naturkunde und Naturschutz, v. 7, p. 245–252.

Betts, S., 2006, Are British bats at risk from wind farms?: British Wildlife, v. 17, p. 339–345.

Brinkman, R., 2006, Survey of possible operational impacts on bats by wind facilities in southern Germany: Gundelfingen, Germany, Report for Administrative District of Freiburg—Department 56, Conservation and Landscape Management, Ecological Consultancy, accessed December 29, 2011, at, *http://www.batcon.org/windliterature.*

Curry, A., 2009, Deadly flights: Science, v. 325, p. 386–387.

De Jong, J., and Ahlen, I., 1991, Factors affecting the distribution pattern of bats in Uppland, central Sweden: Holarctic Ecology, v. 14, p. 92–96.

Durr, T., 2002, Fledermause als Opfer von Windkraftanlagen in Deutschland: Nyctalus, v. 8, p. 115– 118.

Durr, T., and Bach, L., 2004, Bat deaths and wind turbines—A review of current knowledge, and of the information available in the database for

Germany: Bremer Beitrage fur Naturkunde und Naturschutz, v. 7, p. 253–264.

Ferrer, M., de Lucas, M., Janss, G.F.E., Casado, E., Munoz, A.R., Bechard, M.J., and Calabuig, C.P., 2011, Weak relationship between risk assessment studies and recorded mortality in wind farms: Journal of Applied Ecology, doi: 10.1111/j.136502664.2011.02054.x, accessed December, 2011, at *http://www.cb.iee.unibe.ch/content/e7117/e7118/e8764/e9889/e9893/Ferrer_JAppEco2011.pdf*

Hull, C.L., and Muir, S., 2010, Search areas for monitoring bird and bat carcasses at wind farms using a Monte-Carlo model: Australasian Journal of Environmental Management, v. 17, p. 77–87.

Kiefer, A., Merz, H., Rackow, W., Roer, H., and Schlegel, D., 1995, Bats as traffic casualties in Germany: Myotis, v. 32-33, p. 215–220.

Korner-Nievergelt, F., Korner-Nievergelt, P., Behr, O., Niermann, I., Brinkmann, R., and Hellriegel, B., 2011, A new method to determine bird and bat fatality at wind energy turbines from carcass searches: Wildlife Biology, v. 17, p. 350–363.

Lizarraga, J.L., 2003, Seguimiento de la mortalidad de aves y murcielagos en los parques eolicos de Navarra: Guardabosques, v. 21, p. 22–31.

Long, C.V., Flint, J.A., Lepper, P.A., and Didle, S.A., 2009, Wind turbines and bat mortality: interactions of bat echolocation pulses with moving turbine blades: Proceedings of the Institute of Acoustics, v. 31, p. 185–192.

Long, C.V., Flint, J.A., and Lepper, P.A., 2010a, Insect attraction to wind turbines—Does colour play a role?: European Journal of Wildlife Research, v. 57, p. 323–331.

Long, C.F., Flint, J.A., and Lepper, P.A., 2010b, Wind turbines and bat mortality—Doppler shift profiles and ultrasonic bat-like pulse reflection from moving turbine blades: Journal of Acoustic Society of America, v. 128, p. 2238–2245.

Long, C.V., Flint, J.A., Khairul, M., Bakar, A., and Lepper, P.A., 2010c, Wind turbines and bat mortality—Rotor detectability profiles: Wind Engineering, v. 34, p. 517–530.

Nicholls, B., and Racey, P.A., 2007, Bats avoid radar installations—Could electromagnetic fields deter bats from colliding with wind turbines?: PloS ONE, v. 2, no. 3, p. 1-7, 3297.doi: 10.1371/journal.

Rahmel, U. Bach, L., Brinkmann, R., Dense, C., Limpens, H., Mascher, G., Reichenbach, M., and Roschen, A., 1999, Windkraftplanung und fledermause—Konfliktfelder und hinweise zur erfassungsmethodik: Bremer Beitrage fur Naturkunde und Naturschutz, v. 4, p. 155–161.

Rydell, J., Bach, L., Dubourg-Savage, M., Green, M., Rodrigues, L., and Hedenstrom, A., 2010, Bat mortality at wind turbines in northwestern Europe: Acta Chiropterologica, v. 12, p. 261–274.

Trapp, H., Fabian D., Forster, F., and Zinke, O., 2002, Fledermausverluste in einem windpark der oberlausitz: Naturschutzarbeit in Sachsen, v. 44, p. 53–56.

Unpublished Reports and Theses

Anderson, R.L., Strickland, D., Tom, J., Neumann, N., Erickson, W., Cleckler, J., Mayorga, G., Nuhn, G., Leuders, A., Schnieder, J., Backus, L., Becker, P., and Flagg, N., 2000, Avian monitoring and risk assessment at Tehachapi Pass and San Gorgonio Pass wind resource areas, California—Phase 1 preliminary results, *in* National Avian-Wind Power Planning Meeting, 3, San Diego, Calif.,May, 1998, Proceedings: Washington D.C., National Wind Coordinating Committee, p. 31–46.

Anderson, R., Morrison, M.., Sinclair, K., and Strickland, D., 1999, Studying wind energy-bird interactions—A guidance document: Report prepared for the avian subcommittee and the National Wind Coordinating Committee, accessed December 29, 2011, at *http://www.nationalwind.org /publications/wildlife/avian99/Avian_booklet.pdf.*

Anderson, R., Neuman, N., Tom, J., Erickson, W.P., Strickland, M.D., Bourassa, M., Bay, K.J., and Sernka, K.J., 2004, Avian monitoring and risk assessment at the Tehachapi Pass Wind Resource Area—Period of performance—October 2, 1996–May 27, 1998: Golden, Colo., Energy Laboratory, NREL/SR-500-36416, 90 p.

Anderson, R., Neuman, N., Tom, J., Erickson, W.P., Strickland, M.D., Bourassa, M., Bay, K.J., and Sernka, K.J., 2005, Avian monitoring and risk assessment at the San Gorgonio Wind Resource Area, period of performance, March 3, 1997–August 11, 2000: Golden, Colo., Energy Laboratory, NREL/SR-500-38054. 102 pp.

Arnett, E.B., ed., 2005, Relationships between bats and wind turbines in Pennsylvania and West Virginia—An assessment of bat fatality search protocols, patterns of fatality, and behavioral interactions with wind turbines: Final report submitted to Austin, Tex., The Bats and Wind

Energy Cooperative, Bat Conservation International, accessed December 29, 2011, at *http://www.batcon.org/windliterature.*

Arnett, E.B., Hayes, J.P., and Huso, M.M.P., 2006, Patterns of pre-construction bat activity at a proposed wind facility in south-central Pennsylvania: Annual report submitted to Austin, Tex., The Bats and Wind Energy Cooperative, Bat Conservation International, accessed December 29, 2011, at *http://www.batcon.org/windliterature.*

Arnett, E.B., Huso, M.M.P., Reynolds, D.S., and Schirmacher, M., 2007a, Patterns of pre-construction bat activity at a proposed wind facility in northwest Massachusetts: Annual report submitted to Austin, Tex., The Bats and Wind Energy Cooperative, Bat Conservation International.

Arnett, E.B., Schirmacher, M.R., Huso, M.M.P. Huso, and Hayes, J.P., 2009a, Patterns of bat fatality at the Casselman Wind Project in south-central Pennsylvania: Annual report submitted to Austin, Tex., The Bats and Wind Energy Cooperative and the Pennsylvania Game Commission, Bat Conservation International.

Arnett, E.B., Schirmacher, M., Huso, M.M.P., and Hayes, J.P., 2009b, Effectiveness of changing wind turbine cut-in speed to reduce bat fatalities at wind facilities: An annual report submitted to Austin, Tex., The Bats and Wind Energy Cooperative, Bat Conservation International.

Arnett, E.B., Huso, M.M.P., Hayes, J.P., and Schirmacher, M., 2010, Effectiveness of changing wind turbine cut-in speed to reduce bat fatalities at wind facilities: Final report submitted to Austin, Tex., The Bats and Wind Energy Cooperative, Bat Conservation International.

Arnett, E.B., Hein, C.D., Schirmacher, M.R., Baker, M., Huso, M.M.P., and Szewczak, J.M., 2011, Evaluating the effectiveness of an ultrasonic acoustic deterrent for reducing bat fatalities at wind turbines: Final report submitted to Austin, Tex., The Bats and Wind Energy Cooperative, Bat Conservation International.

Baerwald, E., 2007a, Bat fatalities in southern Alberta, *in* Wildlife Research Meeting, 6, San Antonio, Tex., November 2006, Proceedings: Washington D.C., National Wind Coordinating Collaborative. Baerwald, E., 2007b, Prairie winds—Migrating bats and wind energy in Canada: BATS, A publication of Bat Conservation International, v. 25, p. 1–4.

Baerwald, E.F., 2008, Variation in the activity and fatality of migratory bats at wind-energy facilities in southern Alberta—Causes and consequences: Calgary, Alberta, Canada, University of Calgary, Unpublished Master's Thesis. 125 pp.

BHE Environmental, Inc., 2009a, Chiropteran risk assessment: proposed Hardin County North Wind Energy Generation Facility, Hardin County, Ohio: A report prepared for JW Great Lakes Wind, LLC, Cleveland, Ohio.

BHE Environmental, Inc., 2009b, Interim summary: bat acoustic monitoring at the proposed Blue Creek Wind Farm, Paulding and Van Wert Counties, Ohio: A report prepared for Iberdrola Renewables, Inc., Radnor, Pa. [Prepared by BHE, Cincinnati, Ohio, October, 2009.]

BHE Environmental, Inc., 2010a, Chiropteran risk assessment for proposed wind energy generation facility: Hog Creek Wind Farm (I & II), Hardin County, Ohio: A report prepared for Hog Creek Wind Farm, LLC, Cleveland, Ohio. [Prepared by BHE, Cincinnati, Ohio, June, 2010.]

BHE Environmental, Inc., 2010b, Post-construction bird and bat mortality study: Cedar Ridge Wind Farm, Fond Du Lac County, Wisconsin: Interim report prepared for Wisconsin Power and Light, Madison, Wisc. [Prepared by BHE Environmental, Inc., Cincinnati, Ohio, February, 2010.]

Brown, W.K., and Hamilton, B.L., 2002, Draft report—Bird and bat interactions with wind turbines Castle River Wind Farm, Alberta: Report for VisionQuest Windelectric, Inc., Calgary, Alberta, Canada.

Brown, W.K., and Hamilton, B.L., 2006, Bird and bat monitoring at the McBride Lake Wind Farm, Alberta, 2003–2004: Report for Vision Quest Windelectric, Inc., Calgary, Alberta, Canada, accessed December 29, 2011, at *http://www.batcon.org/windliterature.*

Brown, W.K., and Hamilton, B.L., 2006, Monitoring of bird and bat collisions with wind turbines at the Summerview Wind Power Project, Alberta, 2005–2006: Report for Vision Quest Windelectric, Inc., Calgary, Alberta, Canada, accessed December 29, 2011, at *http://www.batcon.org/wind literature.*

California Energy Commission and California Department of Fish and Game, 2007, California guidelines for reducing impacts to birds and bats from wind energy development: Commission Final Report, California Energy Commission, Renewables Committee, and Energy Facilities Siting Division, and California Department of Fish and Game, Resources Management and Policy Division, CEC-700-2007-008-CMF.

Chatfield, A., Erickson, W., and Bay, K., 2009, Avian and bat fatality study, Dillon Wind-Energy Facility, Riverside County, California—Final report—March 26, 2008–March 26, 2009: Prepared for Iberdrola Renewables, Portland, Oreg. [Prepared by Western EcoSystems Technology, Inc., Cheyenne, Wyo., June 3, 2009.]

Curry, R.C., and Kerlinger, P., 2000, Avian mitigation plan—Kenetech model wind turbines, Altamont Pass WRA, California, *in* National Avian Wind Power Planning Meeting, San Diego, Calif., May 1998, Proceedings: Prepared for the Avian Subcommittee of the National Wind Coordinating Committee by LGL Ltd., King City, Ontario, p. 18–27

Derby, C., Dahl, A., Erickson, W., Bay, K., and Hoban, J., 2007, Post-construction monitoring report for avian and bat mortality at the NPPD Ainsworth Wind Farm: Report prepared for Nebraska Public Power District, Western Ecosystems Technology, Inc., Cheyenne, Wyo.

Ecology and Environment, 2009, Bat mist-netting survey report for Black Fork Wind, LLC, Crawford and Richland Counties, Ohio: Prepared for Black Fork Wind, LLC, Denver, Colo. [Prepared by Ecology and Environment, Inc., Overland Park, Kans., October 2009.]

Enk, T., Bay, K., Sonnenberg, M., Baker, J., Kesterke, M., Boehrs, J., and Palochak, A., 2010, Biglow Canyon Wind Farm Phase I post-construction avian and bat monitoring second annual report, Sherman County, Oregon—January 26, 2009–December 11, 2009: Prepared for Portland General Electric Company, Portland, Oreg., [Prepared by Western EcoSystems Technology, Inc., Cheyenne, Wyo., and Walla Walla, Wash., April, 2010.]

Erickson, W.P., Johnson, G.D., Strickland, M.D., and Kronner, K., 2000, Final Report—Avian and bat mortality associated with the Vansycle wind project, Umatilla County, Oregon 1999 study year: Report for Umatilla County Department of Resource Services and Development, Western Ecosystems Technology, Inc., Cheyenne, Wyo.

Erickson, W.P., Johnson, G.D., Strickland, M.D., Young, D.P., Sernka, K.J., and Good, R.E., 2001, Avian collisions with wind turbines—A summary of existing studies and comparison to other sources of avian collision mortality in the United States: Washington, D.C., National Wind Coordinating Committee.

Erickson, W., Johnson, G., Young, D., Strickland, D., Good, R., Bourassa, M., Bay, K., and Sernka, K., 2002, Synthesis and comparison of baseline avian and bat use, raptor nesting and mortality information from proposed and existing wind developments: Portland, Oreg., Bonnevile Power Administration.

Erickson, W.P., Gritski, B., and Kronner, K., 2003, Nine canyon wind power project avian and bat monitoring annual report: Cheyenne, Wyo., Western Ecosystems Technology, Inc., Technical Report. Erickson, W.P., Jeffrey, J., Kronner, K., and Bay, K., 2004, Stateline wind project wildlife monitoring— Final Report, July, 2001 to December, 2003: Western Ecosystem Technology, Inc. and Northwest Wildlife Consultants, Inc., Cheyenne, Wyoming and Pendleton, Oregon.

Erickson, W.P., Jeffrey, J., and Poulton, V.K., 2008, Avian and bat monitoring—Year 1 report: Puget Sound Energy Wild Horse Wind Project, Kittitas County, Washington: Prepared for Puget Sound Energy, Ellensburg, Wash. [Prepared by Western EcoSystems Technology, Inc., Cheyenne, Wyo., January, 2008.]

Exo, K.M., Huppop, O., and Garthe, S., 2003, Birds and offshore wind facilities—A hot topic in marine ecology: Water Study Group Bulletin, v. 100, p. 50–53.

Fiedler, J.K., 2004, Assessment of bat mortality and activity at Buffalo Mountain Windfarm, eastern Tennessee: Unpublished Master's Thesis, University of Tennessee, Knoxville, Tenn.,

Fiedler, J.K., Henry, T.H., Nicholson, C.P., and Tankersley, R.D., 2007, Results of bat and bird mortality monitoring at the expanded Buffalo Mountain Wind Farm, 2005: Knoxville, Tenn., Tennessee Valley Authority.

Good, R.E., Ritzert, M., and Bay, K., 2009, Wildlife baseline studies for the Timber Road Wind Resource Area, Paulding County, Ohio—Final report—September 2, 2008–August 19, 2009: Prepared for Horizon Wind Energy, Houston, Tex. [Prepared by Western EcoSystems Technology, Inc., Bloomington, Ind., December 3, 2009.]

Good, R., Erickson, W., Merrill, A., Simon, S., Murray, K., Bay, K., and Fritchman, C., 2011, Bat monitoring studies at the Fowler Ridge Wind Energy Facility Benton County, Indiana: A report prepared for Fowler Ridge Wind Farm, Western EcoSystems Technology, Inc. [2003 Central Avenue, Cheyenne, Wyo.]

Goodale, W., and Divoll, T., 2009, Birds, bats, and coastal wind farm development in Maine—A literature review: Report BRI 2009-18, Gorham, Maine, BioDiversity Research Institute.

Gritski, B., Kronner, K., and Downes, S., 2008, Leaning Juniper Wind Power Project, 2006–2008— Wildlife monitoring final report: Prepared for PacifiCorp Energy, Portland, Oreg. [Prepared by Northwest Wildlife Consultants, Inc. (NWC), Pendleton, Oreg., December 30, 2008.]

Gritski, B., Downes, S., and Kronner, K., 2009, Klondike III (Phase 1) Wind Power Project wildlife monitoring year one summary, October 2007–October 2008: Prepared for Iberdrola Renewables, Portland, Oreg.

[Prepared by Northwest Wildlife Consultants, Inc. (NWC), Pendleton, Oreg., April 3, 2009.]

Gruver, J.C., 2002. Assessment of bat community structure and roosting habitat preferences for the hoary bat (*Lasiurus cinereus*) near Foote Creek Rim, Wyoming: Laramie, Wyo., University of Wyoming, Unpublished Master's Thesis. 164 pp.

Gruver, J.C., Sonnenburg, M., Bay, K., and Erickson, W., 2008, Post-construction bat and bird fatality study—Blue Sky Green Field Wind Resource Area, Fond du Lac County, Wisconsin: Report prepared for WE Energies, Western Ecosystems Technology, Inc., Cheyenne, Wyo.

Gruver, J.C., Sonnenberg, M., Bay, K., and Erickson, W.P., 2009, Results of a post-construction bat and bird fatality study at Blue Sky Green Field Energy Center, Fond Du Lac County, Wisconsin, July 2008–May 2009: Final report prepared for WE Energies, Milwaukee, Wisconsin, Western Ecosystems Technology, Inc., Cheyenne, Wyo., PSC: 126370.

Hayes, J.P. and Waldien, D.L., 2000, Potential influences of the proposed Condon Wind Project on bats: Unpublished report prepared for CH2MHILL, Portland, Oreg.

Hayes, J.P., and Waldien, D.L., 2000, Potential influences of the Stateline wind project on bats: Unpublished report prepared for CH2MHILL, Portland, Oreg.

Hein, C.D., Rodman, L.B., Schwab, N.A., and Mabee, T.J., 2010, An acoustic study of bat activity at the proposed Coyote Crest Wind Power Project, Washington, Spring–Fall 2008: Unpublished report prepared for Tetra Tech EC, Inc., Portland, Oreg.

Hein, C.D., Arnett, E.B., Schirmacher, M.R., Huso, M.M.P., and Reynolds, D.S., 2011, Patterns of pre-construction bat activity at the proposed Hoosac Wind Energy Project, Massachusetts, 2006–2007: Final project report submitted to the Bats and Wind Energy Cooperative, Bat Conservation International, Austin, Tex.

Hein, C.D., Schirmacher, M.R., Arnett, E.B., and Huso, M.M.P., 2011, Patterns of pre-construction bat activity at the proposed Resolute Wind Energy Project, Wyoming, 2009–2010: Final project report submitted to the Bats and Wind Energy Cooperative, Bat Conservation International, Austin, Tex.

Higgins, K.F., Osborn, R.G., Dieter, C.D., and Usgaard, R.E., 1996, Monitoring of seasonal bird activity and mortality at the Buffalo Ridge Wind Resource Area, Minnesota, 1994–1995—Completion report for the research period May 1, 1994–December 31, 1995: Unpublished report prepared for Kenetech Windpower, Inc. [Prepared by the South Dakota Cooperative Fish and Wildlife Research Unit, Brookings, S. Dak.]

Howe, R.W., Evans, W., and Wolf, A.T., 2002, Effects of wind turbines on birds and bats in northeastern Wisconsin: Madison, Wisc., Wisconsin Public Service Corporation.

Howell, J.A., 1995, Perching prevention assessment at Kenetech 56–100 wind turbine towers: Unpublished report for Kenetech Windpower, San Francisco, Calif.

Howell, J.A., and Didonato, J.E., 1991, Assessment of avian use and mortality related to wind turbine operations, Altamont Pass, Alameda and Contra Costa Counties, California, September 1998 through August 1989: Final report submitted to U.S. Windpower, Inc.

Jacques Whitford, Stantec Limited (Jacques Whitford), 2009, Ripley Wind Power Project postconstruction monitoring report. Project No. 1037529.01—Report to Suncor Energy Products Inc., Calgary, Alberta, and Acciona Energy Products Inc., Calgary, Alberta: Prepared for the Ripley Wind Power Project Post-Construction Monitoring Program. [Prepared by Jacques Whitford, Markham, Ontario.]

Jain, A.A., 2005, Bird and bat behavior and mortality at a northern Iowa windfarm: Ames, Iowa, Iowa State University, Unpublished Master's Thesis. 112 pp.

Jain, A., Kerlinger, P., Curry, R., and Slobodnik, L., 2007, Annual report for the Maple Ridge wind power project post-construction bird and bat fatality study—2006: PPM Energy, Horizon Energy, Curry and Kerlinger, LLC, Cape May Point, N.J.

Jain, A., Kerlinger, P., Curry, R., and Slobodnik, L., 2008, Annual report for the Maple Ridge Wind Power Project: post-construction bird and bat fatality study—2007: Final report prepared for PPM Energy and Horizon Energy and Technical Advisory Committee (TAC) for the Maple Ridge Project Study.

Jain, A., Kerlinger, P., Curry, R., and Slobodnik, L., 2009a, Annual report for the Maple Ridge wind power project post-construction bird and bat fatality study—2007: Annual report prepared for PPM Energy and Horizon Energy. [Prepared by Curry and Kerlinger LLC, Cape May Point, N.J.]

Jain, A., Kerlinger, P., Curry, R., and Slobodnik, L., 2009b, Annual report for the Maple Ridge wind power project post-construction bird and bat fatality study—2008: Annual report prepared for PPM Energy and

Horizon Energy. [Prepared by Curry and Kerlinger LLC, Cape May Point, N.J.]

Jain, A., Kerlinger, P., Curry, R., Slobodnik, L., Fuerst, A., and Hansen, C., 2009c, Annual report for the Noble Ellenburg Windpark, LLC, Post-construction bird and bat fatality study—2008: Prepared for Noble Environmental Power, LLC. [Prepared by Curry and Kerlinger, LLC.]

Jain, A., Kerlinger, P., Curry, R., Slobodnik, L., Histed, J., and Meacham, J., 2009d, Annual report for the Noble Clinton Windpark, LLC, Post-construction bird and bat fatality study—2008: Prepared for Noble Environmental Power, LLC. [Prepared by Curry and Kerlinger, LLC.]

Jain, A., Kerlinger, P., Curry, R., Slobodnik, L., Quant, J., and Pursell, D., 2009e, Annual report for the Noble Bliss Windpark, LLC, Post-construction bird and bat fatality study—2008: Prepared for Noble Environmental Power, LLC. [Prepared by Curry and Kerlinger, LLC.] April 13, 2009.

Jain, A., Kerlinger, P., Slobodnik, L., Curry, R., Fuerst, A., and Harte, A., 2010a, Annual report for the Noble Bliss Windpark, LLC, Post-construction bird and bat fatality study—2009: Prepared for Noble Environmental Power, LLC. [Prepared by Curry and Kerlinger, LLC, Cape May, New Jersey.]

Jain, A., Kerlinger, P., Slobodnik, L., Curry, R., and Russell, K., 2010b, Annual report for the Noble Clinton Windpark, LLC, Post-construction bird and bat fatality study—2009: Prepared for Noble Environmental Power, LLC. [Prepared by Curry and Kerlinger, LLC, Cape May, New Jersey.]

Jain, A., Kerlinger, P., Slobodnik, L., Curry, R., and Russell, K., 2010c, Annual report for the Noble Ellenburg Windpark, LLC, Post-construction bird and bat fatality study—2009: Prepared for Noble Environmental Power, LLC. [Prepared by Curry and Kerlinger, LLC, Cape May, New Jersey.]

James, R.D., 2002, Pickering Wind Turbine, Bird monitoring program in 2002: Report to Ontario Power Generation.

Jeffrey, J.D., Bay, K., Erickson, W.P., Sonneberg, M., Baker, J., Kesterke, M., Boehrs, J., and Palochak, A., 2009, Portland General Electric Biglow Canyon Wind Farm Phase I post-construction avian and bat monitoring first annual report, Sherman County, Oregon, January 2008–December 2008: Technical report prepared for Portland General Electric Company, Portland, Oreg. [Prepared by Western EcoSystems Technology, Inc, Cheyenne, Wyo., and Walla Walla, Wash.]

Johnson, G.D., Young, D.P., Jr., Derby, C.E., Erickson, W.P., Strickland, M.D., and Kern, J.W., 2000a, Wildlife monitoring studies, SeaWest windpower plant, Carbon County, Wyoming, 1995–1999: Cheyenne, Wyo., Western Ecosystems Technology, Inc.

Johnson, G.D., Young, D.P., Jr., Erickson, W.P., Strickland, M.D., Good, R.E., and Becker, P., 2000b, Avian and bat mortality associated with the initial phase of the Foote Creek Rim Wind power Project, Carbon County, Wyoming—November 3, 1998–October 31, 1999: Technical Report prepared for SeaWest Energy Corporation and Bureau of Land Management, 32 p.

Johnson, G.D., Erickson, W.P., Strickland, M.D., Shepherd, M.F., and Shepherd, D.A., 2000c, Avian monitoring studies at the Buffalo Ridge Wind Resource Area, Minnesota—Results of a 4-year study: Technical report prepared for Northern States Power Co., Minneapolis, Minn.

Johnson, G.D., and Strickland, M.D., 2003, Biological assessment for the federally endangered Indiana bat (*Myotis sodalis*) and Virginia big-eared bat (*Corynorhinus townsendii virginianus*), NedPower Mount Storm Wind Project, Grant County, West Virginia: Unpublished report prepared for NedPower Mount Storm, Chantilly, Va. [Prepared by WEST, Inc.]

Johnson, G.D., Erickson, W.P., and White, J., 2003a, Avian and bat mortality at Klondike, Oregon phase I wind plant, Sherman County, Oregon: Cheyenne, Wyo.,Western Ecosystems Technology, Inc.

Johnson, G.D., Perlik, M.K., Erickson, W.P., Strickland, M.D., Shepherd, D.A., and Sutherland, P., Jr., 2003b, Bat interactions with wind turbines at the Buffalo Ridge, Minnesota Wind Resource Area—An assessment of bat activity, species composition, and collision mortality: Electric Power Research Institute, Palo Alto, Calif., and Xcel Energy, Minneapolis, Minn., Electric Power Research Institute Report # 1009178.

Johnson, G.D., and Erickson, W.P., 2010, Avian, bat and habitat cumulative impacts associated with wind energy development in the Columbia Plateau Ecoregion of eastern Washington and Oregon: Final report prepared for Klickitat County Planning Department, Goldendale, Wash. [Prepared by Western EcoSystems Technology, Inc., Cheyenne, Wyo.]

Keeley, B., Ugoretz, S., and Strickland, D., 2001, Bat ecology and wind turbine considerations, *in* Schwartz, S.S., ed., National Avian Wind Power Planning Meeting, 6, Carmel, Calif., May 16–17, 2000, Proceedings: Washington D.C., National Wind Coordinating Collaborative, p. 135–146.

Kerlinger, P., 2002, An assessment of the impacts of Green Mountain Power Corporation's wind power facility on breeding and migrating birds in

Seasburg, Vermont: Unpublished report for the Vermont Department of Public Service, Montpelier, Vt.

Kerlinger, P., Curry, R., and Ryder, R., 2000, Ponnequin wind energy project avian studies, Weld County, Colorado—Summary of activities during 2000: Prepared for Public Service Company of Colorado, Denver, Colo.

Kerlinger, P., Curry, R., Culp, L, Jain, A., Wilkerson, C., Fischer, B., and Hasch, A., 2006, Post-construction avian and bat fatality monitoring study for the High Winds wind power project, Solano County, California: McLean, N.J., Curry and Kerlinger, LLC, Two-year report.

Kerlinger P., Curry, R., Culp, L., Hasch, A., and Guarnaccia, J., 2007, Migratory bird and bat study at the Crescent Ridge Wind Power Project, Bureau County, Illinois—September 2005–August 2006: Final Draft prepared for Orrick, Herrington, & Sutcliffe, LLP, Curry and Kerlinger, LLC, McLean, Va.

Kerns, J., and Kerlinger, P., 2004, A study of bird and bat collision fatalities at the MWEC Wind Energy Center, Tucker County, West Virginia: Cape May Point, N.J., Curry and Kerlinger, LLC, Annual report for 2003.

Kerns, J., Erickson, W.P., and Arnett, E.B., 2005, Bat and bird fatality at wind-energy facilities in Pennsylvania and West Virginia, *in* Arnett, E.B., ed., Relationships between bats and wind turbines in Pennsylvania and West Virginia—An assessment of bat fatality search protocols, patterns of fatality, and behavioral interactions with wind turbines: Final report submitted to the Bats and Wind Energy Cooperative, Bat Conservation International, Austin, Tex., p. 24–95.

Koford, R., Jain, A., Zenner, G., and Hancock, A., 2004, Avian mortality associated with the Top of Iowa Wind Farm—Progress Report, Calendar Year 2003: Ames, Iowa, Iowa State University, Iowa Cooperative Fish and Wildlife Research Unit.

Krenz, J. D., and McMillan, B.R., 2000, Final teport—Wind-turbine related bat mortality in southwestern Minnesota: St. Paul, Minn., Minnesota Department of Natural Resources.

Mabee, T.J., Cooper, B.A., and Plissner, J.H., 2004, A radar study of nocturnal bird migration at the proposed Mount Storm wind power development, West Virginia, Fall 2003: Unpublished report prepared by ABR, Inc.. Fairbanks, Alaska. [Prepared for WEST, Inc. and Nedpower.]

Miller, A., 2008, Patterns of avian and bat mortality at a utility-scaled wind farm on the southern high plains: Lubbock, Tex., Texas Tech University, Unpublished Master's Thesis.

Miller, A., Boal, C., Nagy, L., and Woeck, B., 2009, Estimating avian and bat mortality and identifying spatial and temporal distribution at a utility-scale wind energy development, *in* National Wind Coordinating Collaborative (NWCC) Wildlife and Wind Research Meeting, 7, Milwaukee, Wisc., October 28–29, 2008 (Pre-Conference Session, October 27, 2008), Proceedings: Prepared for the National Wind Coordinating Collaborative by S.S. Schwartz.

Morrison, M., 2002, Searcher bias and scavenging rates in bird-wind energy studies: Unpublished report prepared for the National Renewable Energy Laboratory, NREL/SR-500-30876, Golden, Colo. National Research Council, 2007, Environmental impacts of wind-energy projects: Washington, D.C., National Academies Press.

Nicholson, C.P., 2001, Buffalo Mountain Windfarm bird and bat mortality monitoring report—October 2000–September 2001: Knoxville, Tenn., Tennessee Valley Authority.

Nicholson, C.P., 2003, Buffalo Mountain wind facility bird and bat mortality monitoring report— October 2001 to September 2002: Knoxville, Tenn., Tennessee Valley Authority.

Nicholson, C.P., Tankersley, J.R.D., Fiedler, J.K., and Nicholas, N.S., 2005, Assessment and prediction of bird and bat mortality at wind-energy facilities in the southeastern United States: Knoxville, Tenn., Tennessee Valley Authority, Final Report.

Northwest Wildlife Consultants, Inc. and Western EcoSystems Technology, Inc., 2007, Avian and bat monitoring report for the Klondike II Wind Power Project, Sherman County, Oregon: Prepared for PPM Energy, Portland, Oreg. Managed and conducted by NWC, Pendleton, Oreg. Analysis conducted by WEST, Cheyenne, Wyo.

Orloff, S., and Flannery, A., 1992, Wind turbine effects on avian activity, habitat use, and mortality in Altamont Pass and Solano County Wind Resource Areas, 1989–1991: Sacramento, Calif.,California Energy Commission.

Piorkowski, M.D., 2001, Breeding bird habitat use and turbine collisions of birds and bats located at a wind farm in Oklahoma mixed-grass: Stillwater, Okla., Oklahoma State University, Unpublished Master's Thesis.

Poulton, V., and Erickson, W., 2010, Post-construction bat and bird fatality study, Judith Gap Wind Farm, Wheatland County, Montana: Final report prepared for Judith Gap Energy LLC, Western Ecosystems Technology, Cheyenne, Wyo.

Puzen, S.C., 2002, Bat interactions with wind turbines in northeastern Wisconsin: Green Bay, Wisc., Wisconsin Public Service Corporation of Green Bay, Wisconsin.

Redell, D., Arnett, E.B., Hayes, J.P., and Huso, M., 2006, Patterns of pre-construction bat activity at a proposed wind facility in south-central Wisconsin: Final report submitted to the Bats and Wind Energy Cooperative, Bat Conservation International, Austin, Tex.

Schmidt, E., Piaggio, A.J., Bock, C.E., and Armstrong, D.M., 2003, National Wind Technology Center site environmental assessment—Bird and bat use and fatalities: Golden, Colo., National Renewable Energy Laboratory, Final report NREL/SR-500-32981.

Stantec Consulting, Inc. (Stantec), 2009a, Post-construction monitoring at the Mars Hill Wind Farm, Maine—Year 2, 2008: Prepared for First Wind Management, LLC, Portland, Maine. [Prepared by Stantec Consulting, Topsham, Maine.]

Stantec Consulting, Inc. (Stantec), 2009b, Spring, summer, and fall 2008 bird and bat survey report for the Buckeye Wind Facility, Ohio: Prepared for EverPowerWind Holdings, Inc. [Prepared by Stantec Consulting, formerly Woodlot Alternatives, Inc., Topsham, Maine.]

Stantec Consulting, Inc. (Stantec), 2009c, Stetson I Mountain Wind Project, Year 1 post-construction monitoring report, 2009 for the Stetson Mountain Wind Project in Penobscot and Washington Counties, Maine: Prepared for First Wind Management, LLC, Portland, Maine. [Prepared by Stantec, Topsham, Maine.]

Stantec Consulting Inc. (Stantec), 2008a, 2007 spring, summer, and fall post-construction bird and bat mortality study at the Mars Hill Wind Farm, Maine: Prepared for UPC Wind Management, LLC, Cumberland, Maine. [Prepared by Stantec Consulting, formerly Woodlot Alternatives, Inc., Topsham, Maine.]

Stantec Consulting Inc. (Stantec), 2008b, Post-construction monitoring at the Munnsville Wind Farm, New York—2008: Prepared for EON Climate and Renewables, Austin, Tex. [Prepared by Stantec Consulting, Topsham, Maine.]

Stantec Consulting Ltd. (Stantec Ltd.), 2010, Wolfe Island Ecopower Centre post-construction followup plan—Bird and bat resources monitoring report No. 2—July–December 2009, File No. 160960494: Prepared for TransAlta Corporation's wholly owned subsidiary, Canadian Renewable Energy Corporation. [Prepared by Stantec Ltd., Guelph, Ontario, Canada.]

Stantec Consulting Services Inc. (Stantec), 2010, Cohocton and Dutch Hill Wind Farms year 1 post-construction monitoring report, 2009, for the Cohocton and Dutch Hill Wind Farms in Cohocton, New York: Prepared for Canandaigua Power Partners, LLC, and Canandaigua Power Partners II, LLC, Portland, Maine. [Prepared by Stantec, Topsham, Maine.]

Stantec Consulting Services, Inc. (Stantec), 2010, Bird and bat pre-construction surveys for Kingdom Community Wind Project in Lowell, Vermont: Unpublished report prepared for Green Mountain Power, Colchester, Vt. [Prepared by Stantec Consulting, South Burlington, Vt.]

Strickland, M.D., Arnett, E.B., Erickson, W.P., Johnson, D.H., Johnson, G.D., Morrison, M.L., Shaffer, J.A., and Warren-Hicks, W., 2011, Comprehensive guide to studying wind energy/wildlife interactions: Prepared for the National Wind Coordinating Collaborative, Washington, D.C.

Tennessee Valley Authority, 2003, Draft environmental assessment—20-MW windfarm and associated energy storage facility: Knoxville, Tenn., Tennessee Valley Authority.

Thelander, C.G. and Rugge, L., 2000, Bird risk behaviors and fatalities at the Altamont Wind Resource Area, *in* National Avian Wind Power Planning Meeting, 3, San Diego, Calif., May 1998, Proceedings: Washington, D.C. National Wind Coordinating Committee/RESOLVE, p. 5–14.

Tierney, R., 2007, Buffalo Gap I Wind Farm avian mortality study—February 2006–January 2007— Final survey report: Prepared for AES SeaWest, Inc. TRC, Albuquerque, N. Mex., TRC Report No. 110766-C-01.

Tierney, R., 2009, Buffalo Gap 2 Wind Farm Avian Mortality Study—July 2007–December 2008 Final study report: Prepared for Buffalo Gap Wind Farm 2 LLC, TRC, Albuquerque, N. Mex.

TRC Environmental Corporation, 2008, Post-construction avian and bat fatality monitoring and grassland bird displacement surveys at the Judith Gap Wind Energy Project, Wheatland County, Montana: Prepared for Judith Gap Energy, LLC, Chicago, Ill., TRC Project 51883-01 (112416). [Prepared by TRC Environmental Corporation, Laramie, Wyo.]

Tucker, V.A., 1996, Using a collision model to design safer wind turbine rotors for birds: Durham, N.C., Duke University, Department of Zoology, accessed on March 28, 2012, at *http://www.batsandwind.org/pdf/Using%20a%20collision%20model%20to%20design%20safer%20turbines%20for%20birds.pdf.*

Erickson, W.P., and Sharp, L., 2005, Phase 1 and Phase 1A avian mortality monitoring report for 2004– 2005 for the SMUD Solano Wind Project:

Prepared for Sacramento Municipal Utility District (SMUD), Sacramento, Calif.

URS Corporation, 2010a, Final Marengo I Wind Project year one avian mortality monitoring report: Prepared for PacifiCorp, Salt Lake City, Utah. [Prepared by URS Corporation, Seattle, Wash.]

URS Corporation, 2010b, Final Marengo II Wind Project year one avian mortality monitoring report: Prepared for PacifiCorp, Salt Lake City, Utah. [Prepared by URS Corporation, Seattle, Wash.]

U.S. Department of Energy, 2002, Final site-wide environmental assessment of National Renewable Energy Laboratory's National Wind Technology Center: Golden, Colo., National Renewable Energy Laboratory, U.S. Department of Energy.

Wind Turbine Guidelines Advisory Committee, 2010, Consensus recommendations on developing effective measures to mitigate impacts to wildlife and their habitats related to land-based wind-energy facilities: Prepared for the U.S. Department of the Interior, Washington, D.C. [Prepared by Kearns and West.]

Young, D.P., Jr., Erickson, W.P., Strickland, M.D., and Good, R.E., 2002, Comparison of avian effects from UV light reflective paint applied to wind turbines—Foote Creek Rim Wind Plant, Carbon County, Wyoming: Golden, Colo., National Renewable Energy Laboratory.

Young, D.P., Jr., Erickson, W.P., Good, R.E., Strickland, M.D., and Johnson, G.D., 2003, Avian and bat mortality associated with the initial phase of the Foote Creek Rim wind power project, Carbon County, Wyoming—November 1998 to June 2002: Cheyenne, Wyo., Western Ecosystems Technology, Inc.

Young, D.P., Jr., Erickson, W.P., Jeffrey, J.D., Bay, K.J., Good, R.E., and Lack, B.G., 2005, Avian and sensitive species baseline study plan and final report Eurus Combine Hills Turbine Ranch Umatilla County, Oregon: Cheyenne, Wyo., Western Ecosystems Technology, Inc.

Young, D.P,. Jr., Jeffrey, J., Erickson, W.P., Bay, K., and Poulton, V.K., 2006, Eurus Combine Hills Turbine Ranch Phase 1 post construction wildlife monitoring first annual report: Technical report prepared for Eurus Energy America Corporation, San Diego, Calif., and the Combine Hills Technical Advisory Committee, Umatilla County, Oreg. [Prepared by Western EcoSystems Technology, Inc., Cheyenne, Wyo., and Northwest Wildlife Consultants, Inc., Pendleton, Oreg.]

Young, D.P., Jr., Erickson, W.P., Jeffrey, J., and Poulton, V.K., 2007, Puget Sound Energy Hopkins Ridge Wind Project Phase 1 post-construction avian and bat monitoring first annual report, January– December 2006: Technical report for Puget Sound Energy, Dayton, Washington and Hopkins Ridge Wind Project Technical Advisory Committee, Columbia County, Wash. Prepared by Western EcoSystems Technology, Inc., Cheyenne, Wyoming, and Walla Walla, Washington.

Young, D.P., Jr., Erickson, W.P., Bay, K., Nomani, S., and Tidbar, W., 2009, Mount Storm Wind Energy Facility, Phase 1 post-construction avian and bat monitoring: Report prepared for NedPower Mount Storm LLC. [Prepared by Western Ecosystems Technology, Inc., Cheyenne, Wyo.]

Young, D.P., Jr., Bay, K., Nomani, S., and Tidhar, W.L., 2010, Nedpower Mount Storm Wind Energy Facility, Post-Construction Avian and Bat Monitoring—March–October 2009: Prepared for NedPower Mount Storm, LLC, Houston, Tex. [Prepared by Western EcoSystems Technology, Inc., Cheyenne, Wyo.]

Young, D.P., Jr., Nomani, S., Tidhar, W.L., and Bay, K., 2011, NedPower Mount Storm Wind Energy Facility post-construction avian and bat monitoring, July–October 2010: Unpublished report prepared for NedPower Mount Storm, LLC, Houston, Tex. [Prepared by Western EcoSystems Technology, Inc., Cheyenne, Wyo.]

Wind Resources

American Wind Energy Association, 2012, American Wind Energy Association, accessed on January 14, 2012, at http://www.awea.org/learnabout/industry_stats/index.cfm.

Bats and Wind Energy Cooperative, 2012, Bats and Wind Energy Cooperative, accessed on January 14, 2012, at http://www.batsandwind.org/.

Canadian Wind Energy Association, 2012, Canadian Wind Energy Association, accessed on January 14, 2012, at http://www.canwea.ca/images/uploads/File/NRCan_-_Fact_Sheets/canwea-factsheeteconomic-web.pdf.

Gipe, P., 2004, Wind power—Renewable energy for home, farm, and business (2d ed): Chelsea Green, White River Junction, Vt.

Pasqualetti, M., Richter, R., and Gipe, P., 2004, History of wind energy, in Cleveland, C.J., ed., Encyclopedia of energy, vol. 6: San Diego, Calif., Academic Press, p. 419–433.

Redlinger, R.Y., Dannemand, P., and Morthorst, P.E., 2002, Wind energy in the 21st century: New York, New York, Palgrave Macmillan.

World Wind Energy Report, 2010, World Wind Energy Report, accessed January 14, 2012, at http://www.wwindea.org/home/images/stories/pdfs/worldwindenergyreport2010_s.pdf.

In: Bats and Wind Energy
Editor: Sean Willis

ISBN: 978-1-63483-616-6

Chapter 2

PRELIMINARY METHODOLOGY TO ASSESS THE NATIONAL AND REGIONAL IMPACT OF U.S. WIND ENERGY DEVELOPMENT ON BIRDS AND BATS*

J. E. Diffendorfer, J. A. Beston, M. D. Merrill, J. C. Stanton, M. D. Corum, S. R. Loss, W. E. Thogmartin, D. H. Johnson, R. A. Erickson and K. W. Heist

CONVERSION FACTORS

Multiply	By	To obtain
	Length	
meter (m)	3.281	foot (ft)
	Area	
square kilometer (km^2)	0.3861	square mile (mi^2)
	Power	
megawatt (MW)	1×10^6	joules per second (J/s)
gigawatt (GW)	1×10^9	joules per second (J/s)
	Energy	
megawatthour (MWh)	3.6×10^9	joule (J)

* This is an edited, reformatted and augmented version of U.S. Geological Survey Scientific Investigations Report 2015–5066, issued May 2015.

ABBREVIATIONS, ACRONYMS, AND SYMBOLS

a	age at first reproduction
BBS	North American Breeding Bird Survey
c	annual chance an individual will die from a collision with a wind turbine
$CV(N)$	coefficient of variation of population size
DOI	U.S. Department of the Interior
EIA	U.S. Energy Information Administration
EIS	environmental impact statement
F	recovery factor
FRI	fatality-risk index
FT	proportion of annual fatalities caused by turbines
GAMMS	Guidelines for Assessing Marine Mammal Stocks
GW	gigawatt
h	number of habitats used by a species
IEA	International Energy Agency
IRI	indirect-risk index
IUCN	International Union for Conservation of Nature
K	carrying capacity
m	maternity
MARSS	multivariate autoregressive state-space (model)
MMPA	Marine Mammal Protection Act
MW	megawatt
MWh	megawatthour
N	population size
N_{min}	lower bound on an estimate of population size
N_t	population size at year t
n	annual fatalities from wind turbines
NA	not applicable
NABat	North American Bat Monitoring Program
NEPA	National Environmental Policy Act
NREL	National Renewable Energy Laboratory
p	percentage of a species' range that overlaps with the locations of wind turbines
PBR	potential biological removal
R_2	coefficient of determination
r_{max}	maximum annual population growth rate under optimal conditions

RR	risk ratio
s	adult survival
SGCN	species of greatest conservation need
SWAP	State wildlife action plan
U.S.C.	U.S. Code
USFWS	U.S. Fish and Wildlife Service
USGS	U.S. Geological Survey
WEIAM	Wind Energy Impacts Assessment Methodology (USGS project)
λ	population growth rate
λ_s	growth rate of the baseline population, with no fatalities from wind energy facilities
λ_w	growth rate of a population, with fatalities from wind energy facilities

ABSTRACT

The U.S. Geological Survey has developed a methodology to assess the impacts of wind energy development on wildlife; it is a probabilistic, quantitative assessment methodology that can communicate to decision makers and the public the magnitude of these effects on species populations. The methodology is currently applicable to birds and bats, focuses primarily on the effects of collisions, and can be applied to any species that breeds in, migrates through, or otherwise uses any part of the United States. The methodology is intended to assess species at the national scale and is fundamentally different from existing methods focusing on impacts at individual facilities.

Publicly available fatality information, population estimates, species range maps, turbine location data, biological characteristics, and generic population models are used to generate both a ranked list of species based on relative risk as well as quantitative measures of the magnitude of the effect on species' population trend and size. Three metrics are combined to determine direct and indirect relative risk to populations. A generic population model is used to estimate the expected change in population trend and includes additive mortality from collisions with wind turbines. Lastly, the methodology uses observed fatalities and an estimate of potential biological removal to assess the risk of a decline in population size. Data for six bird species have been processed through the entire methodology as a test case, and the results are presented in this report.

Components of the methodology are based on simplifying assumptions and require information that, for many species, may be sparse or unreliable. These assumptions are presented in the report and should be carefully considered when using output from the methodology. In addition, this methodology can be used to recommend species for more intensive demographic modeling or highlight those species that may not require any additional protection because effects of wind energy development on their populations are projected to be small.

1.0. Introduction

Recent growth in wind energy generation has led to concerns over the effect of this development on wildlife in the United States. Investigations of impacts to volant species (winged species capable of flying) are conducted at many wind energy facilities, yet there remains a paucity of knowledge regarding the effects on species at the national and regional level (Arnett and others, 2008; Katzner and others, 2013). The U.S. Geological Survey (USGS) is using research, monitoring data, and modeling to develop a probabilistic and quantitative methodology to assess both the current and future population-level consequences of wind energy development on those species of birds and bats that are present in the United States during any part of their life cycle. The methodology is national in scope because the population assessed is defined as all the individuals of a species that occur in the United States, not a subset of them in a particular region or State. The methodology specifically addresses whether the fatalities caused by collisions with wind turbines can be sustained by a species in the long term. It also produces a ranked list of species' potential responses to the habitat loss associated with wind energy facilities. The methodology also uses projections of new wind energy in 2025 to make model projections about future effects of wind energy development on birds and bats. The USGS developed this methodology in response to renewable energy initiatives of the U.S. Department of the Interior (DOI), starting with the New Energy Frontier initiative (U.S. Department of the Interior and U.S. Department of Agriculture, 2011) and continuing with the Powering Our Future initiative (U.S. Department of the Interior, 2014).

USGS scientists studying minerals and energy have decades of experience in producing transparent, thoroughly vetted methods to assess (1) undiscovered resources of conventional and continuous oil and gas, coal, gas hydrates, and minerals and (2) potential geologic carbon dioxide storage resources. Geographers, ecologists, and statisticians of the USGS conduct a wide array of applied research on species population biology and spatial distributions, with an emphasis on responses to stressors at multiple scales. This assessment methodology is a product of the interdisciplinary cooperation of these scientists, and devising a methodology is the first step toward completing a full assessment. The objectives of this report are to describe the methodology, its central question and key assumptions, and how it would be implemented. A case study of six bird species is provided as an example of implementation and output.

The methodology was developed over a 3-year period. Initially, the authors participated in meetings with stakeholders, including a large kickoff meeting and targeted meetings with industry representatives, conservation organizations, and individual Federal agencies involved in energy and wildlife issues related to wind energy development. From these meetings, three key issues became apparent that influenced the goals for the methodology. First, the stakeholders felt it was most important to understand the population-level consequences of wind energy development on species. Second, they wanted a rapid method to prioritize species in terms of their risk from wind energy generation. Third, they were concerned about both the direct effects of wind energy development caused by collisions and the indirect effects caused by habitat loss and behavioral avoidance of wind turbines. Once the general goals of the methodology were defined, the components of the methodology were developed, tested using data from actual species, and discussed with species experts, all of which led to refinement and the current methodology.

Wildlife populations can be described by their size and trend, and thus "population-level consequences" can be defined as changes in a population's size or trend. The determinants of a population's size and trend are numerous and varied and likely differ both across species and across different areas or time periods for the same species. Information on these determinants is unavailable for most species, and overly complex models that require assumptions about much of the input may not provide better population analyses than simplified models (Morris and Doak, 2002).

This methodology uses different approaches to measure potential population-level consequences from wind energy facilities. All methods focus on flexibility and applicability to multiple species without attempting to describe the nuances of complex population dynamics. This focus may undermine the accuracy of some estimates, but the estimates will be useful for determining which species are likely to experience population-level consequences from wind energy development and for addressing whether the fatalities caused by collisions with wind turbines can be sustained by a species.

The methodology is not intended to supplant the more detailed analyses required to make decisions under the Endangered Species Act (16 U.S.C. § 1531 et seq.), the Bald and Golden Eagle Protection Act (16 U.S.C. § 668 et seq.), or the Migratory Bird Treaty Act (16 U.S.C. §§ 703–712).

1.1. State of Wind Energy Development

Wind-powered electricity generation has increased significantly over the last decade to 167 million megawatthours (MWh) in 2013, which represents a cumulative installed capacity of 62.3 gigawatts (GW) in the United States by September 2014 (American Wind Energy Association, 2014). Wind energy generation currently represents 31.4 percent of U.S. electricity from renewable sources and 4.1 percent of total net electricity generation (U.S. Energy Information Administration, 2014b, Table 7.2a). Wind energy is growing at a rapid pace and has overtaken all but conventional hydroelectric generation for renewable energy sources (Figure 1). Projections for U.S. wind energy generation by 2025, from reports by the U.S. Energy Information Administration (2014a) and the International Energy Agency (2012), suggest that installed capacity could nearly double to a range from 80 to 114 GW (see section 3.0 of this report).

1.2. Research on the Impact of Wind Energy Development on Wildlife

Academic and government scientists, including those from the USGS (Phillips, 2011; Ellison, 2012), have conducted research on the effects on wildlife from wind energy development. Electricity is generated from wind-driven turbines, and the physical collision of the turbine blades with volant species, such as birds and bats, causes injury and death. Fatalities from collisions are considered "direct effects" throughout this report. This report includes the potential for barotrauma (damage to body tissue caused by the difference in air pressure around the turbine blade) in the potential for collision damage (Baerwald and others, 2008; Grodsky and others, 2011; Rollins and others, 2012). The construction of wind energy facilities, the road networks required to service them, and the energy grid necessary to transport electricity to consumers can also affect wildlife through habitat loss and alteration;

moreover, some species avoid areas near turbines (Stevens and others, 2013; Winder and others, in press). Habitat loss and alteration, as well as avoidance behavior, are considered indirect effects in this report. This assessment methodology primarily addresses direct effects, but the prioritization approach (section 2.4) includes an output ranking of species' potential responses to indirect effects.

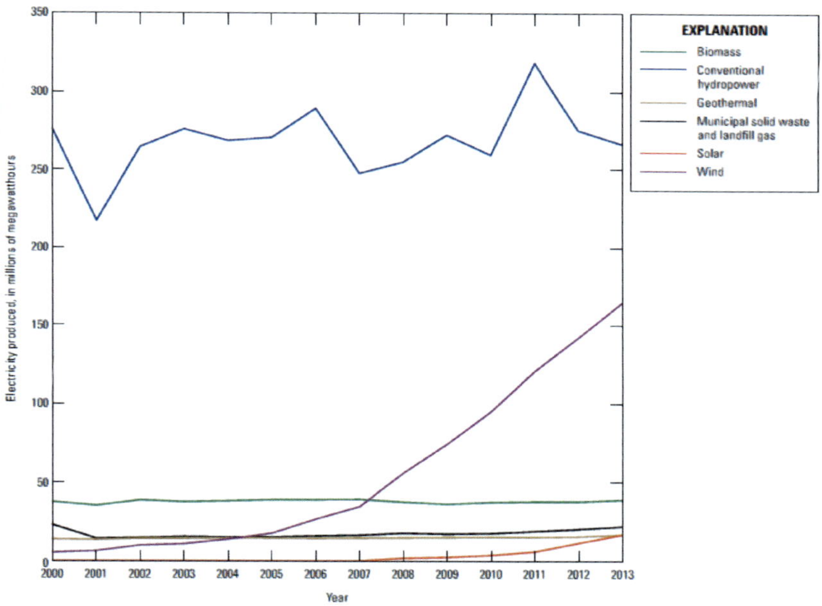

Figure 1. Graph showing U.S. electricity production from renewable energy sources, in millions of megawatthours, from 2000 through 2013. Wind energy use in this time interval increased more than the use of other renewable sources. Data from U.S. Energy Information Administration (2014b, Table 7.2a).

Review papers such as those by W.P. Erickson and others (2001, 2014), National Research Council, Committee on Environmental Impacts of Wind-Energy Projects (2007), Arnett and others (2008), and Strickland and others (2011) provide thorough summaries of existing research on the effects of wind energy facilities on wildlife and introduce the large and active research community working on this issue in the United States. Works by international authors and agencies or groups show that the effects of these facilities also are of concern outside the United States; see studies by the United Kingdom Joint Nature Conservation Committee (Crockford, 1992), Scottish Natural Heritage

(Gill and others, 1996), Council of Europe (Langston and Pullan, 2004), Canadian Wildlife Service (Kingsley and Whittam, 2005), Rydell and others (2010), and Voigt and others (2012).

The scientific methods for assessing the effects of wind energy facilities on the population size and trend (the increasing, decreasing, or stable pattern of the population through time) of species over regional or larger areas are still developing. Conducting an assessment on species over large areas can have considerable temporal and logistical constraints. Each individual species can be the subject of a highly complex multiyear effort; examples include research on whooping cranes (Pearse and Selbo, 2012; Butler and others, 2013) and golden eagles (U.S. Fish and Wildlife Service, 2011a, b; Pagel and others, 2013; Johnston and others, 2014; Watson and others, 2014). Likewise, each wind energy facility can be monitored and assessed for years, but such intensive efforts, although useful for a few species of greatest concern, are impractical for studying all species and all locations.

A few attempts have been made to estimate the population-level consequences on bird and bat populations of fatalities caused by collisions with wind turbines. Although the result is not a direct estimate of population-level effects, W.P. Erickson and others (2014) divided an estimate of the annual fatalities per year by a population estimate for 20 bird species calculated by the Partners in Flight Science Committee (2013). Carrete and others (2009) developed a spatially explicit population model for Egyptian vultures and investigated the role of increased fatalities from wind turbines on population trends. Bellebaum and others (2013) modeled annual fatalities of 3.1 percent of the population for red kites in Germany and used the potential biological removal (PBR) method (see section 2.4) to estimate a PBR value of 4 percent, indicating the species may be near an unsustainable level of fatality. In another study of red kites, Schaub (2012) used an individual-based computer simulation of individual kites to study how the location of turbines might affect kite population trends.

When estimating effects at the population level is not possible, prioritizing species by using qualitative approaches can assist decision making that is related to wind energy generation and wildlife. In several studies, researchers developed methods for prioritizing birds relative to their potential risk from wind turbines. These studies targeted marine birds that could be affected by offshore wind energy development. Garthe and Hüppop (2004) used nine factors (such as flight maneuverability, adult survival, and current conservation status) to develop a wind facility sensitivity index. Desholm (2009) ranked 38 marine bird species in terms of their relative abundance and the elasticity of population growth to adult survival. Bright and others (2008) developed a basic sensitivity ranking (high versus medium) for birds in Scotland by using variables associated with sensitivity to wind facilities obtained from a literature review and combined these values with data from other sources to develop an overall risk map for wind energy facilities in Scotland.

1.3. The Assessment Methodology

The common principles underpinning the USGS methodologies currently in use for assessing undiscovered oil and gas and geologic carbon dioxide storage resources have been adopted in this methodology for assessing the effects of wind energy development on wildlife (Schmoker and Klett, 2005; Brennan and others, 2010; Charpentier and Cook, 2011). Methodologies for assessing resources (geological, hydrological, biological, and so on) normally include two components. First is the methodology's step-by-step procedure for estimating and using input parameters to calculate an output specific to the resource question. Second is the methodology's implementation plan, a structured approach that focuses on maintaining consistency and ensuring that the steps of the methodology are followed with peer oversight. The USGS Wind Energy Impacts Assessment Methodology (WEIAM) project team is currently tasked only with the completion of the first component, the authoring of the steps and calculations of the methodology. Although the implementation procedure is still in the initial stages of development, it is anticipated that it will follow established USGS procedures. According to procedure, a core group of scientists (generally the creators of the methodology plus other experts, as needed) serve as the assessment panel. An assessor (most likely a conservation ecologist, biologist, or biostatistician) will give multiple presentations to the panel during the course of an assessment. These presentations (1) establish the need to assess a particular species and (2) allow the panel to review the input data that the assessor plans to use in the various assessment components of the methodology (discussed further in section 2). The panel then implements the models and calculations of the methodology to produce outputs.

1.3.1. Principles

The USGS methodology described in this report is designed to be quantitative, with results reported in a probabilistic manner, and to be completely transparent in process and structure. Similar concepts in conservation biology and applied ecological modeling also have been used in ecological risk assessment (Athreya and Karlin, 1971; Boyce, 1992). Quantitative results are the most useful output for an assessment. Numerical results can be analyzed more readily than qualitative results, generally provide a clearer statement of finding, and can best inform decisions. A quantitative result is only as valuable as the certainty associated with that value; therefore, the methodology described here, like others produced by the USGS, is probabilistic, producing low and high bounds around a most likely result. Higher levels of uncertainty result in a greater spread between upper and lower bounds.

A transparent process helps maintain the consistent application of the method and allows researchers external to the USGS to conduct assessments using the same methods and vet the results. An additional value of this transparency is the ease of improvement through iterative upgrades to the methodology. Methodologies often are revised and modified over time as knowledge and technology improve. For example, the USGS has published numerous revisions to its assessment methodology for continuous (unconventional) undiscovered oil and gas. These changes were adopted as significant increases in production of continuous resources and additional data revealed limitations in previous methods. Thus, as studies of the effects of wind energy facilities on birds and bats continue, we anticipate changes in knowledge that will affect the methodology. Because this is the first version of a methodology to assess the effects of wind energy development on wildlife at broad scales, it could certainly be improved through future advances in research and understanding. Ongoing research by the WEIAM project team that may lead to improvements of the methodology is included in appendix 1.

Finally, in a complex multipart methodology such as the one described here, the various components should be structured and linked to each other to create a cohesive methodology that achieves the designers' goal. For example, the results for each component should be both internally consistent and consistent across components. Furthermore, individual components should produce unbiased results in the face of incomplete or biased input data. Lastly, because science activities related to the management of species often intersect with policy, structuring the methodology to avoid subjective policy decisions is necessary. When avoidance is not possible, the methodology must clearly demarcate when policy issues affect the structure of the methodology and how these nonscientific decisions will be included in a functioning assessment.

1.3.2. Assessment Unit

In general terms, the assessment unit describes the individual entity that the methodology assesses; it is the core of the assessment process. The results from an assessment are calculated by using the methodology and are presented at the assessment-unit level. The methodology described here works with U.S. populations of individual species and will apply to any species that is present in the United States during any part of its life cycle. For birds and bats, species designations follow those in the "American Ornithologists' Union Checklist of North and Middle American Birds" (American Ornithologists' Union, 2014) and "Mammal Species of the World" (Wilson and Reeder, 2005), respectively.

The spatial scale and population assessed will change from species to species, though the methodology will focus on the population of individuals that intersect with the United States. For example, a nonmigratory species distributed in the southwestern United States and Mexico will be assessed using that portion of the population that resides in the United States. For migratory species, the portion of the total population that may be present in the United States at any time during its life cycle will be assessed. Thus, if all individuals of a neotropical migrant bird species may be found in the United States during a stopover while flying between their more northern breeding habitat and more southern overwintering grounds, this entire population will be assessed because all of the individuals move through the United States annually and are possibly exposed to wind turbines in the United States. In other words, the assessment considers the impacts of all the wind turbines that exist in a species' range within the United States, including Alaska and Hawaii. Smaller regions within a species' range that may reflect demographic or genetic subunits could be assessed with the same methodology if interest, need, and data were available. However, the methodology is not designed to estimate site- or facility-specific impacts of wind energy development on a local population.

1.3.3. Output and Potential Uses

The assessment methodology produces outputs to address two main objectives. First, it produces ranked lists of species based on potential direct risk (from collisions) and indirect risk (from habitat change) from wind energy development. This prioritization meets the objective to quickly filter bird and bat species with respect to their risk by ranking them into qualitative relative risk categories.

Second, for those species prioritized as high *direct* risk, the methodology produces two outputs that address the objective of quantifying population-level consequences from collision fatality. The first is an estimate of the expected change in population trend with the addition of turbine collision mortality based on a demographic population model. The second is an estimate of the risk ratio, which quantifies whether the fatalities from collision with wind turbines would reduce the population below a target size. Thus, the demographic model approaches population-level consequences by estimating changes in trend, while the risk ratio addresses population-level consequences resulting in changes in population size.

1.3.4. Intended Audience

USGS assessment results, and the methodologies that produce them, are created with multiple end users in mind. Due to the national scale and numerous assessment units involved, USGS national assessments are typically of most use to decision makers, government agencies, and U.S. citizens who need a broad, generalized understanding of a particular resource. That understanding is based on the summation of individual quantitative results for each assessment unit, but these results should not be confused with site-specific or single-facility impact assessments. Assessments such as environmental impact statements (EIS) that are completed to comply with the National Environmental Policy Act (NEPA) (42 U.S.C. § 4321 et seq.) are commonly site specific. The method described here also is not related to the siting guidelines established by the U.S. Fish and Wildlife Service (USFWS) for reducing potential wildlife impacts at specific wind energy facilities (U.S. Fish and Wildlife Service, 2012).

2.0. METHODOLOGY

2.1. Methodology Overview

The methodology will be implemented through a formal assessment process based on those currently in use by the energy and mineral programs at the USGS. A generalized flowchart of the methodology is presented in Figure 2; detailed flowcharts of the components of the process are provided in the applicable sections of this report. For each species, an assessor will gather data (step 1 in Figure 2) and develop a species description that (1) summarizes any studies related to the species' responses to wind turbines, including studies related to the species' response to habitat loss, fragmentation, and roads; (2) includes the sources and reasoning for the model parameter estimates and their distributions, the observed fatality estimates at turbines, any time series of abundance data, and estimates of population size; (3) synthesizes information about the species to define the population being assessed and its spatial scale; and (4) includes a species distribution map. This information will be presented in both oral and written form to a peer-review assessment panel and modified as warranted.

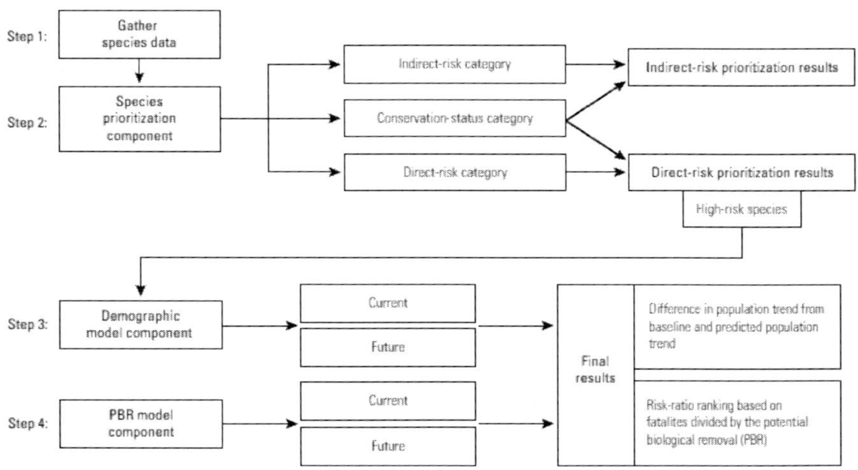

Figure 2. Flowchart showing the generalized steps in the assessment methodology. "Current" refers to methodology steps based on installed capacity and number of turbines in 2014. "Future" refers to steps that use projections of installed capacity and number of turbines for 2025.

Once data are collected, species will be prioritized with respect to direct risk and indirect risk from wind facilities (step 2 in Figure 2). Direct-risk prioritization will be based on two metrics that estimate a species' potential risk from the fatalities caused by collisions with turbines: the proportion of annual fatalities due to turbines (FT) (section 2.2.3.1) and the fatality-risk index (FRI) (section 2.2.3.2). The values of FT and FRI will be combined with a species' conservation status to rank species on the basis of their potential risk from collisions with turbines (Table 1). Species classified into a high-risk category during an external review of the risk rankings (explained below in section 2.2.3) will then be further assessed by using the demographic model and risk ratio. Indirect-risk prioritization will be based on the indirect-risk index (IRI), which will measure the potential consequences of behavioral avoidance, habitat loss, and degradation associated with wind energy facilities (see section 2.2.4). The IRI value will be combined with a species' conservation status to rank species on the basis of their potential risk from habitat modification (Table 1).

For prioritization, most efforts will be spent amassing species-level information and developing distribution or range maps from available sources such as eBird (Cornell Lab of Ornithology, 2015) or the Global Biodiversity Information Facility Secretariat (2015). This work also could include a formal process for eliciting expert opinion for those species lacking data (Runge and others, 2011; Martin and others, 2012).

Prioritization will be conducted for all bird and bat species that occur in the United States during any part of their life cycle.

Once prioritization is complete, the species ranked highly for direct risk will be assessed by using the demographic model (step 3 in Figure 2) and the risk ratio (step 4 in Figure 2). The demographic model estimates a change in population trend caused by fatalities from collision with wind turbines. All estimates are probabilistic and include measures of uncertainty. For each species, estimating the population trend may require estimating demographic rates from raw data, performing a meta-analysis from existing studies, or again eliciting expert opinion. The risk ratio estimates how close the estimated annual fatalities from wind turbines are to the potential biological removal (PBR), which estimates total fatalities that can occur before the population will decline below a target population size. Prioritization will be done relatively rapidly across many species simultaneously; however the demographic model and PBR-based risk ratio will require more effort and will rely on the panel process described above.

Late in the demographic-model and the PBR-based risk-ratio processes (steps 3 and 4 in Figure 2), the respective models are repeated using estimates of projected wind energy development, described in section 3, for the highly ranked species. Thus, prioritization is only done using current levels of wind energy development, and then the highly ranked species are assessed using the demographic model and the risk ratio for the current levels and two projected levels of wind energy development. Species will not be reprioritized under projected levels of wind energy development because their future conservation status cannot be reliably predicted.

Table 1. Ranking system for species analysis in the prioritization component of the assessment methodology based on combinations of conservation status and direct or indirect risk

Rank	Direct- or indirect-riskcategory	Conservation-status category
1	High	High
2	High	Medium
3	Medium	High
4	High	Low
5	Medium	Medium
6	Low	High
7	Medium	Low
8	Low	Medium
9	Low	Low

The final output of the assessment includes the following information for each species: (1) its relative risk of response to habitat loss or degradation (indirect-risk prioritization) and (2) its relative risk of population-level consequences from wind energy development (direct-risk prioritization). In the direct-risk category, for species having a high prioritization rank, the final output also includes (1) the estimated change (with uncertainty) in population trend (demographic model) and (2) the risk ratio, which is based on the estimated annual fatalities and the PBR.

2.2. Species Prioritization

The first component of the assessment methodology, species prioritization, is designed to rapidly characterize a large number of bird and bat species in terms of their relative risk from wind energy facilities. International conservation organizations, Federal and State agencies, and others involved in resource management commonly prioritize species by using variables such as current population size, population trend, and known threats. The use of prioritization as a tool in species management has been studied by conservation biologists; although uncertainty exists in the prioritization process (Burgman and others, 1999), risk-based ranking approaches can correctly estimate extinction risk with a 70- to 80-percent success rate (Keith and others, 2004).

USGS investigators developed the prioritization component partially on the basis of two earlier studies. Desholm (2009) combined information about exposure (the proportion of a species' total population that moved through the area of a wind energy facility) and the ability of the species to respond to added mortality from wind energy facilities (the elasticity of adult survival), whereas W.P. Erickson and others (2014) examined the mortality rates of small passerines at wind energy facilities by dividing estimates of annual fatalities by continent-wide population estimates. The approach described here expands on these ideas by considering impacts beyond single facilities and including more indicators of a species' response to wind energy.

2.2.1. Implementation

The species prioritization approach uses two general types of information to assess the relative risk of population-level consequences to a species from wind energy facilities: (1) a species' current conservation status and (2) estimates of direct risk and indirect risk. A species' conservation status is likely correlated with its ability to withstand added mortality from wind turbines or habitat loss. For example, imperiled species with higher conservation status should, in general, be more vulnerable to any additional adverse impacts than species with lower conservation status (Andelman and others, 2004). Estimates of direct risk and indirect risk quantify some of the potential impacts of wind energy facilities on a species. The combination of a species' current conservation status and its level of direct or indirect risk determines its ranking relative to other species.

Each species is assigned three risk metrics that combine to determine its rank for direct- and indirect-risk prioritizations (Figure 3): (1) relative risk based on the species' conservation status (section 2.2.2), (2) relative risk from collision fatalities, or "direct risk" (section 2.2.3), and (3) relative risk from habitat modification, or "indirect risk" (section 2.2.4). Metrics 1 and 2 are combined using a qualitative approach to assign a direct-risk ranking between 1 and 9 (Table 1), and metrics 1 and 3 are combined in the same way to assign an indirect-risk ranking (Table 1).

The ranking system shown in Table 1 currently emphasizes direct or indirect risks over conservation status when ranking species. For example, high direct or indirect risk in combination with medium conservation status is ranked higher (rank 2) than its inverse (rank 3). Given the goals of the assessment, this ranking system was established to emphasize the possible effects of wind energy facilities more than a species' current conservation status and is an initial effort that will require review by experts, decision makers, and regulators prior to an implemented assessment.

To rank species, breakpoints that place species into high, medium, or low categories must be set for the metrics that are used to describe conservation status and direct or indirect risk. Because these types of breakpoints are somewhat arbitrary and frequently include both scientific and policy-related considerations, the methodology instead estimates an average rank by randomly sampling breakpoints from across the range of observed values for each metric.

For each set of random breakpoints, species are assigned a numerical rank (1–9) by combining their maximum values for conservation status and either direct or indirect risk; this rank is then averaged across all sets of randomly selected breakpoints. For direct risk, the highest risk category for either the proportion of fatalities due to turbines (FT) or the fatality-risk index (FRI) determines the direct-risk category ranking, and the indirect-risk category is determined by the indirect-risk index (IRI). The final outputs from prioritization are two ranked lists of species: (1) a direct-risk prioritization list with an average rank value for the combination of direct risk and conservation status and (2) an indirect-risk prioritization list with an average rank value for the combination of indirect risk and conservation status. A species with limited data may appear on only one of the lists until more information is available.

A key decision point would occur after prioritization. Given a list of averaged direct-risk prioritization scores for each species, a subset of these must be selected for assessment using the demographic model and risk ratio. This step is ultimately subjective, as how far down the average rankings one chooses to go is not specifically dictated by this methodology. Ultimately, the USGS would not make this decision. Instead, the USGS would convene a stakeholder meeting consisting of members of Federal and State agencies responsible for implementing laws and policies related to wind energy development and wildlife issues to review the prioritization process and its outputs and to determine the cutoff for species that should be further assessed.

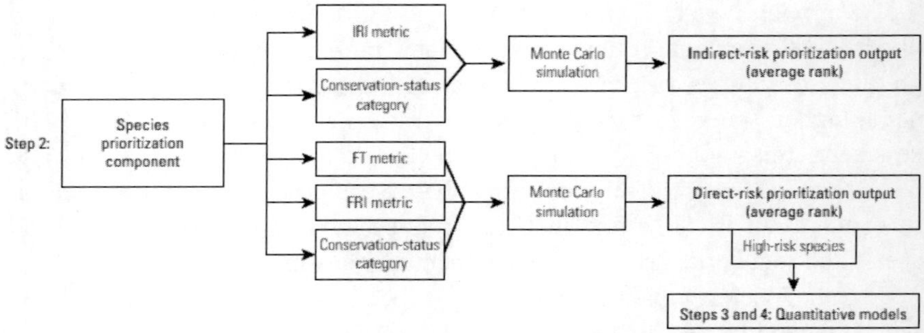

Figure 3. Flowchart of the species prioritization component of the assessment methodology. FRI, fatality-risk index; FT, proportion of annual fatalities due to turbines; IRI, indirect-risk index.

2.2.2. Current Conservation Status

The conservation status of many species in the United States has been classified by a variety of organizations for different purposes and with different standards of assessment. This methodology uses the conservation status measured at the State level by the proportion of States that listed the species as “Species of Greatest Conservation Need” in a State wildlife action plan (SWAP; Beach and others, 2011). Other potential sources of conservation status information include (1) the Red List of the International Union for Conservation of Nature (2012, 2014), (2) listing by the Endangered Species Act of 1973, as amended (16 U.S.C. §1531 et seq.), and (3) the list, “Birds of Conservation Concern 2008” (U.S. Fish and Wildlife Service, 2008).

2.2.3. Direct Risk

2.2.3.1. Proportion of Fatalities Due to Turbines

The most obvious impact of wind energy generation on wildlife is usually fatalities of birds and bats from collisions with turbine blades. Various studies have quantified fatalities at individual wind energy facilities (Smallwood and

Karas, 2009; Kitano and Shiraki, 2013), and researchers have projected fatality rates to larger scales (Loss and others, 2013; Smallwood, 2013). W.P. Erickson and others (2014) divided species-specific fatality estimates by population size to compare mortality rates from turbines across a suite of small passerines. This approach is unsuitable for comparisons across more varied species because of differences in natural history. Instead, the proportion of annual fatalities due to turbines (FT) compares species-specific fatality estimates from wind energy facilities to annual fatalities from all sources.

To calculate FT, the estimate of the number of individuals killed by wind turbines annually (n) is divided by the product of population size (N) and the adult mortality rate (calculated as 1 minus adult survival, s):

$$\mathrm{FT} = \frac{n}{(1-s)N} \tag{1}$$

In general, long-lived species with low mortality rates are more likely to experience additive mortality from anthropogenic sources than short-lived species with inherently high mortality (Péron and others, 2013). Therefore, species with a higher FT value (that is, number of animals killed at wind energy facilities divided by number of animals that die from any cause) are at higher risk of population-level consequences from wind energy facilities. Risk assessment using FT may be biased toward higher values for those species that strongly compensate for turbine mortality with a reduction in mortality from other sources or an increase in reproduction. Thus, FT is a conservative measure of risk.

The number of individuals killed by wind turbines each year, n, could be directly estimated from fatality studies if these data were available. Currently, direct estimates of fatality from turbine collisions across a species' range are not available for any species. Alternatively, one can coarsely estimate n by multiplying the total avian fatalities from wind turbines by the proportion of observed fatalities attributed to each species. These types of estimates have been done for birds (Zimmer-ling and others, 2013; W.P. Erickson and others, 2014), but not for bats, although such estimates for bats are possible. In making these estimates, the researcher assumes that (1) the wind energy facilities chosen for sampling to estimate fatalities are representative of all wind facilities in the United States and (2) the carcasses of species killed by turbine collisions have similar levels of detectability. Both of these assumptions are clearly violated by the available studies, yet no alternative approaches have been developed. Estimating species-specific fatality rates with this approach can produce considerable bias in the assessment output, perhaps leading to unreliable results. Output from the methodology will be improved if species-specific fatality rates become available.

Total population size, *N*, is estimated by a number of organizations for birds, but is very difficult to assess for most species of bats. The Partners in Flight program (Partners in Flight Science Committee, 2013), the Midwinter Waterfowl Survey (Sharp and others, 2002), and the Waterbird Conservation for the Americas program (Kushlan and others, 2002) generate estimates of population size for some, but not all, bird species in the United States. Bat population sizes are difficult to estimate because of their nocturnal behavior, small size, and similar appearances across species, which make them difficult to identify by sight. A few cave-roosting species of bats in the United States are monitored at a sufficient number of caves to produce rough estimates of population size (Thogmartin and others, 2012), but bats currently are not systematically monitored in the United States. The North American Bat Monitoring Program (NABat) seeks to address this data deficiency (U.S. Geological Survey, 2015; Loeb and others, in press). When estimates of population size are not available, rough bounds based on expert judgment may be required (Russell and others, 2014).

Survival has been estimated for birds by a variety of sources and individual studies. These estimates were published in species-specific demographic modeling papers by the Institute for Bird Populations (Michel and others, 2006) and in Birds of North America species accounts (Poole, 2005). Data for bats are sparse, but most bats are long lived and estimates of survival exist for some species (Barclay and Harder, 2003; McCracken, 2003; O'Shea and others, 2003; Frick and others, 2007; Thogmartin and others, 2012). When survival estimates are unavailable for a species, it may be possible to substitute estimates from closely related species. Note that the survival estimate used when calculating adult mortality should include fatalities from wind energy facilities. If not, FT becomes a ratio of wind-facility-related fatality to non-wind-facility-related fatality, rather than the proportion of wind-facility-related fatality to all fatalities. If a survival rate does not include fatalities from wind energy facilities, then the estimated fatalities from wind should be added to $(1 - s)N$ in the denominator of equation 1.

2.2.3.2. Fatality-Risk Index

Annual fatalities from wind turbines are not always available and may be poorly estimated for some species. To account for the lack of data, the methodology uses the fatality-risk index (FRI), which is based on the assumption that species with slow life histories and high exposure to wind turbines are more likely to be at risk of population effects from fatalities caused by collisions. A similar assumption was made in a prioritization approach designed for use at individual wind farms (Desholm, 2009). To calculate the FRI, the percentage of the population impacted (p) is divided by an index of life-history speed (the ratio of maternity [m] to age at first reproduction [a]):

$$\mathrm{FRI} = \frac{p}{m/a} \tag{2}$$

For birds, data for these values are in the same sources as data for survival, whereas for bats, they are more difficult to obtain. For birds, maternity could be taken directly from the literature or calculated as the product of nest success rate, average clutch size, hatchability rate, average number of clutches per year, and a presumed 1:1 sex ratio of offspring.

The ratio of m a was developed to account for the known relation between life-history traits and the elasticity of population trend as related to these demographic parameters (Oli, 2004; Stahl and Oli, 2006). For example, the population trend for a species with a ratio of m a greater than 0.75 is most sensitive to changes in reproductive parameters, whereas the population trend of a species with a ratio of m a less than 0.25 is most sensitive to survival. Because wind turbine collisions affect survival, m a should measure the potential sensitivity of species to turbine collisions. Ultimately, m a is a measure of life-history speed; some species have a low ratio because they have only a few offspring (small m) and delay reproduction (large a), whereas other species have a high ratio because they produce many offspring (large m) and reproduce earlier in their life (often a equals 1). Larger values of the FRI indicate a higher risk, which can be a result of either the presence of many wind turbines in a species' range or the slow life history of the species.

Calculating the proportion of a species' range that overlaps with turbine locations requires both a range map for a species and a map of the known locations of turbines. A range map is a geospatial data layer representing the spatial distribution of a species and may show variation in abundance across space or simply the presence or absence of the species. Turbine locations are available from a national turbine dataset developed by researchers working on this project (Diffendorfer and others, 2014). For both birds and bats, migration makes defining and mapping the range of a species complex because abundance changes across space during different times of the year, and for bats, fatalities increase during migration (Cryan, 2011; Ellison, 2012). Furthermore, some species may move through particular geographic regions intensively during migration (Miller and others, 2014).

For birds, breeding-season ranges and distributions of many species have been mapped with the North American Breeding Bird Survey (BBS) data (Sauer and others, 2015). Maps of overwintering locations could potentially be developed by using data from the Christmas Bird Count (National Audubon Society, 2015). Maps of year-round distribution, including breeding, migratory, and overwintering areas, could perhaps be developed by using eBird, a citizen-science-based repository of bird locations (Cornell Lab of Ornithology, 2015). Ideally, the resulting maps would be used to integrate the proportion of the population near wind energy facilities over an entire year as an estimate of p. Regardless of the final data source(s) used to make distribution maps, the same mapping method must be applied to all species during prioritization to ensure that relative rankings are comparable across species. The BBS abundance maps include an estimate of relative abundance in each grid cell (21.475 square kilometers). The relative abundance in each grid cell in the United States (some species' ranges included Canada) can be multiplied by the area of each grid cell, and then standardized to sum to 1 across all grid cells in the United States. Doing so creates an estimate of the proportion of the species' breeding population in each grid cell across the United States; see an example in Figure 4. The cumulative proportion of the population in grid cells containing one or more turbines is one estimate of p. This approach may overestimate p because isolated turbines may affect only a small portion of the population that actually is located in a grid cell. Conversely, if a grid cell is used as a migratory pathway or wintering habitat, then p may be underestimated when based on data from only breeding birds.

Scientists in North America do not yet have a centralized monitoring program for bats (see NABat at U.S. Geological Survey, 2015, and Loeb and others, in press) or a repository for distribution maps. These species may require distribution modeling or range-map development as part of estimating p, as may some bird species. In addition, because bat fatalities are the highest during migration (Arnett and others, 2008; Cryan, 2011), seasonal range maps might be most useful for modeling p in bats. Appendix 1 describes additional research that the USGS is performing to improve estimates of p through an improved understanding of how animals use airspace.

2.2.4. Indirect-Risk Index

Species that rarely collide with wind turbines can still suffer population consequences due to disturbance, displacement, and habitat fragmentation and loss (Leddy and others, 1999; Langston and Pullan, 2004; Percival, 2005; Fox and others, 2006). Scientists consider indirect effects to be a potentially serious consequence of wind energy. For example, Kuvlesky and others (2007, p. 2490) stated:

> European conservationists generally consider the habitat loss associated with wind farm developments to be a greater threat to bird populations than are collision fatalities.

This sentiment was repeated by Katzner and others (2013). A number of studies suggest that some species avoid turbines at variable distances (Leddy and others, 1999; Stevens and others, 2013). Furthermore, wind energy facilities include roads and transmission lines, both of which can affect species (Forman and Alexander, 1998; Coffin, 2007).

Although the population-level consequences caused by indirect effects have not been quantified for most species, information about the natural history of species may help categorize risk. In general, highly specialized species are more sensitive to changes in habitat than generalist species (Swihart and others, 2003; Munday, 2004); in one study of the indirect effects of turbines on wintering birds, only one of the most specialized species considered (Le Conte's sparrow; *Ammodramus leconteii*) appeared to be displaced by turbines (Stevens and others, 2013). The indirect-risk index (IRI) was developed to take advantage of the expectation that species that use fewer habitats will be more sensitive to the indirect effects of wind energy facilities and is calculated as

$$\mathrm{IRI} = \frac{p}{h} \tag{3}$$

where p, as above, is the percentage of a species' range that overlaps with the turbines and h is the number of habitats used by a species. Information on the number of habitats considered suitable for each species is available from the IUCN's species database (International Union for Conservation of Nature, 2014). Higher values of IRI indicate a higher risk because a large proportion of the population is exposed to turbines or the species occupies fewer habitats.

Because of stakeholders' concerns, the WEIAM project team tried to include a measure of species response to indirect effects from wind energy development; however, the team found that a generalizable and quantitative approach that linked indirect effects to changes in population status was not feasible. In light of that realization, the IRI is estimated and included as part of the indirect-risk prioritization, but it is not used beyond this point in the methodology. Species that rank highly for indirect risk may require further scrutiny and research to understand if indirect effects from wind energy facilities do, in fact, occur.

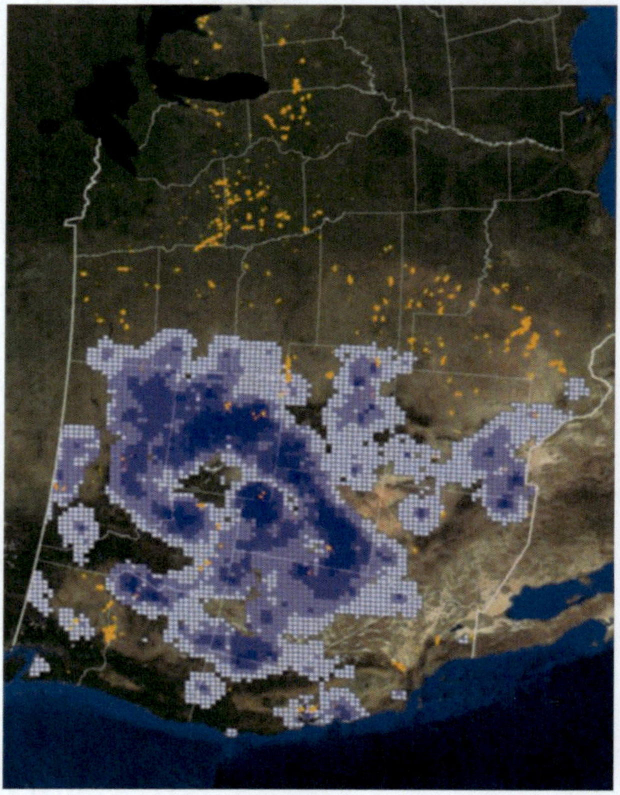

Base imagery from Esri.

Figure 4. Map of the Central and Western United States showing the distribution of golden eagles (purple grids) relative to turbine locations (gold dots). Darker colors in the grid cells represent a high normalized relative abundance. Golden eagle data are from the North American Breeding Bird Survey (Sauer and others, 2015). Turbine locations are from Diffendorfer and others (2014).

2.3. Demographic Model

2.3.1. Background

Demographic models represent a well-developed set of mathematical tools in both population biology and ecological risk assessment. They can be used to understand the interactions between management actions and population trends (Crouse and others, 1987), compare projected population dynamics under different scenarios (Rose and Cowan, 2003), and investigate the factors and processes contributing to observed population trends (Koons and others, 2005; Thogmartin and others, 2013). Methods and tools for implementing demographic models are well established, including commercial software programs and open-source packages (for example, R.A. Erickson and others, 2014). Furthermore, the behavior of demographic models is well understood, and scientists have developed a framework for using them in risk assessments and population viability analyses (Burgman and others, 1993; Beissinger and McCullough, 2002; Morris and Doak, 2002).

The general goal of the modeling used in this assessment is to understand if, and by how much, observed levels of fatalities from wind turbines across a species' range in the United States affect that species' population trend given the uncertainties associated with the number of fatalities, the processes driving the demography of the species, and the observed population trend (when available). Although population trend is not the only variable associated with the status of a population, it does indicate whether a population is growing or declining and if it is correlated with estimates of extinction risk (O'Grady and others, 2004). Developing models of biological phenomena must balance generality, precision, and realism because an ecological model cannot simultaneously include high levels of all of these characteristics (Levins, 1966). For this USGS assessment, a modeling approach should be general enough to apply to many species, including those lacking time series of abundance estimates, yet have sufficient realism and precision to produce results capable of indicating risk to the overall population from wind energy development. However, appendix 1 describes an example of a more complex model developed for Indiana bats that the USGS can use to develop a tailored assessment for *Myotis* bat species in general.

After considering a number of alternative approaches (see appendix 2), the WEIAM team developed a method based on simple, generalized population models to compare population trends with and without the addition of fatalities caused by turbines. This component is termed the "demographic model" throughout this report (Figure 5). The demographic model includes two steps: (1) estimation of population growth rate in the absence of wind energy facilities and (2) estimation of the change in population growth rate with the addition of fatalities from wind energy facilities.

2.3.2. Estimating Population Growth Rate

The population growth rate is the proportional change in population size from one year to the next. Annual growth rates and associated measures of uncertainty can be calculated with a time series of population size estimates (or a time series of population size indices). Monitoring programs, such as the BBS, routinely estimate population trends using such time series, and the current assessment methodology can use a similar approach to directly estimate population trends for species susceptible to direct impacts from wind energy production.

In the absence of empirical estimates of population growth rate, the dominant eigenvalue of a matrix population model is an alternative estimate of population growth rate (Caswell, 2000). Matrix models include assumptions that shape their applicability to particular ecological problems. These assumptions are well described by a number of authors (Ebert, 1998; Caswell, 2000) and include the assumption that animals can be classified by age, density independence, a stable age distribution, and constant vital rates. For this component of the methodology, a stage-structured, postbreeding matrix model, with an annual time step, is used to describe the female segment of a population (Caswell, 2000; Morris and Doak, 2002). The age of first reproduction determines the size (number of rows and columns) of the matrix. For example, species that begin breeding within the first year of life would have a matrix with 2 rows and 2 columns, whereas a matrix of 4 rows by 4 columns would model species that begin breeding in the third year of life. The values in the first row determine recruitment (new individuals entering the population), and the subsequent rows describe annual survival after birth or hatching and each year thereafter. The second-to-last column represents juveniles that survive and mature to reproductive age and then reproduce before the next year.

$$\begin{bmatrix} 0 & 0 & \cdots & s_j m & s_a m \\ s_j & 0 & \cdots & 0 & 0 \\ 0 & s_j & \cdots & 0 & 0 \\ \vdots & \vdots & \ddots & \vdots & \vdots \\ 0 & 0 & \cdots & s_j & s_a \end{bmatrix}$$

Parameters include survival (*s*, the chance a female survives from one year to the next) and maternity (*m*, the number of female offspring per female per year). Thus, fecundity is defined as the female offspring born to females surviving since the previous census ($s_j \times m$ or $s_a \times m$, for juvenile and adult females, respectively). Estimates of survival and maternity are those used in the prioritization component (section 2.2). To parameterize these distributions, means and variances are taken from (in decreasing order of quality) the literature or available data, professional opinion, data on surrogate species, or an assumed value based upon theoretical expectations. For most species, when data on differential breeding success by age or life stage are not available, a single maternity value will be assumed for all reproductive stages.

For both empirical estimates and matrix-based estimates of population growth rate, uncertainty or bias in the estimates of input parameters can introduce bias in the estimated trend. Care must be taken to ensure that estimates are representative of the population occurring in the United States and that they do not include populations already affected by wind energy development. In addition, the estimated population growth rate must be accompanied by some descriptor of the associated uncertainty, which can be estimated directly from time series or propagated through a matrix model by sampling survival from a beta distribution and maternity from a lognormal distribution.

2.3.3. Estimating Change in Population Growth Rate

To estimate the change in population growth caused by collisions with wind turbines, estimates of fatalities must be made and then used to estimate a reduction in the existing population trend. The approach uses the turbine mortality rate (*c*), which is the annual chance an individual will die from a collision with a wind turbine. If wind fatalities are assumed to affect all species' stage and age classes at the same rate, then *c* is the same across all individuals, and an unstructured population model can be used to determine how those fatalities affect the population growth rate.

In the absence of density dependence, populations grow exponentially, and the population size of a species is equal to its population size the previous year multiplied by its population growth rate, λ. Mathematically, population dynamics can be written as $N_{t+1} = \lambda N_t$, where N_t is the population size in year t. With the addition of fatalities from wind, $N_{t+1} = \lambda N_t - c\lambda N_t$, where $c\lambda N_t$ is the number of individuals killed by wind turbines between t and $t+1$. With minor mathematical rearrangement, $N_{t+1} = (1-c)\lambda N_t$ and the growth rate of a population experiencing a turbine mortality rate of c is equal to the population growth rate without fatalities caused by wind turbines multiplied by $(1-c)$. This population growth rate can be written $\lambda_w = (1-c)\lambda_b$, where λ_w is the growth rate of a population including the turbine mortality rate and λ_b is the growth rate of the baseline population, with no fatalities from wind energy facilities. It follows that the change in population growth rate is equal to $c\lambda_b$. To estimate the change in population growth rate, the turbine mortality rate must first be estimated.

Current and projected future (see section 3.0) turbine-collision mortality rate estimates can be calculated by dividing the number of individuals of each species estimated to be killed in turbine collisions each year by the estimated population size. The estimated number of individuals killed by wind turbines each year, n, is described in section 2.2.3.1. For birds, the range of uncertainty in the denominator of the turbine-collision mortality estimate FT (equation 1 above) spans an order of magnitude, stemming from the range used by the Partners in Flight Science Committee (2013) for total U.S. population sizes for each bird species (Rich and others, 2004; Confer and others, 2008). For bats, uncertainty on population size estimates will be extremely large, mainly because population sizes have not been estimated for most species.

To estimate the effects of collisions with wind turbines, random samples of population growth rate without fatalities from turbines are drawn from a normal distribution with mean and standard deviation from estimates in section 2.3.2, and random samples of turbine mortality rate are drawn from a triangular distribution, where the mode is n divided by the Partners in Flight Science Committee's (2013) point estimate of population size and the end points from the order of magnitude uncertainty bounds. The change in population growth rate and resulting expected population growth rate with wind are then calculated for each set of samples in section 2.2.3 above, generating distributions for both the change and the expected growth rate. The methodology then estimates the probability that the population trend is <1 for both the original growth rate and the growth rate with the effects of wind turbines considered. The change between the observed and predicted probability that the population trend is <1 is reported as output. This output, the change in the probability that the population trend is <1, represents an estimate of the added risk that fatalities from wind turbines pose to a species' population trend.

In the demographic model component, collision fatalities are assumed to be additive to natural mortality, and awareness of this assumption is critical to interpreting the results. In reality, many species are likely to compensate for wind mortality by means of reductions in other sources of mortality or increased reproduction, and thus the estimates of change in trend are most likely overestimates. Thus, the results from this component should be considered estimates of the maximum possible change in population trend caused by the fatalities. The role of additive versus compensatory mortality on population dynamics is not well understood (Burnham and Anderson, 1984; Nichols and others, 1984); therefore, developing stochastic simulation models that included density dependence and compensation was not feasible in the generalized approach (appendix 2). The potential biological removal component, described next, is an alternative approach that attempts to include the role of density dependence while addressing collision fatalities.

2.4. Potential Biological Removal and Risk Ratio

2.4.1. Background

Potential biological removal (PBR) estimates indicate the total number of animals that could be killed as marine bycatch before a population would decline below a population size deemed sustainable, often considered half of a species' carrying capacity (Wade, 1998). Since its inception, using the PBR has become a standard approach for managing human-induced deaths of marine mammal species (Taylor and others, 2000), and scientists have studied various details of the approach, such as how it is used to predict risk (Wade, 1998; Milner-Gulland and Akçakaya, 2001), how parameters can be estimated (Niel and Lebreton, 2005; Dillingham and Fletcher, 2008), and how management decisions and assumptions of risk influence PBR assessments (Lonergan, 2011; Moore and others, 2013). In addition, the PBR and variations of it have been applied to bird species in relation to deaths caused by long-line fishing (Richard and Abraham, 2013), hunting (Runge and others,

2009), and wind turbines (Bellebaum and others, 2013). The PBR is calculated as

$$PBR = F\frac{r_{max}}{2}N_{min} \tag{4}$$

where r_{max} is the maximum annual population growth rate under optimal conditions, N_{min} is a lower bound on an estimate of the population size, and F is a recovery factor, set by decision makers. In the PBR, $r_{max}/2$ represents the rate of take (number of individuals harvested, or killed) maximizing the net productivity of a population when logistic growth is assumed (Wade, 1998). The parameter F is essentially a safety factor set by decision makers (1) to adjust the value of the PBR to increase the rate of recovery of populations that are too small or (2) to account for uncertainties in the data used to calculate the PBR and to ensure that the PBR is not set too high. Under the Marine Mammal Protection Act (MMPA; 16 U.S.C. § 1362(20)), values for F can fall between 0.1 and 1.

In practice, estimated values of the PBR are compared to observed fatality rates to gage the level of risk for a species. If fatalities are well below the PBR, then a population should remain above an acceptable level and be at low risk. In a study by Richard and Abraham (2013), the ratio of fatalities to the PBR was calculated by using Monte Carlo approaches and was called the "relative risk"; this ratio is called the risk ratio (RR) in the USGS methodology. When the value for the RR is much less than 1, fatalities are much lower than the PBR and a species is at low risk. As RR values increase, risk increases, and at values at or above 1, the PBR is either met or exceeded by fatalities and the species is at high risk.

Currently, both the National Marine Fisheries Service and the U.S. Fish and Wildlife Service calculate PBRs for marine mammal stocks, including cetaceans, pinnipeds, sea otters (*Enhydra lutris*), polar bears (*Ursus maritimus*), and West Indian or American manatees (*Trichecus manatus*). They follow the regularly updated "Guidelines for Assessing Marine Mammal Stocks" (GAMMS; National Marine Fisheries Service, 2015) to perform these assessments. This methodology adopts these guidelines because of their historical and current use by the Federal Government and the levels of research and effort used to regularly update them. For each species with sufficient data, the PBR is calculated by using a Monte Carlo approach that includes the uncertainty associated with r_{max}.

The maximum rate of population growth under optimal conditions, r_{max} , is difficult to estimate because, for most species, populations are rarely observed existing in optimal conditions. Estimates of r_{max} are available for a few species, such as bacteria in petri dishes, species expanding into new areas, or those recovering from a population crash or overharvest (Gedamke and others, 2009). A number of approaches exist to estimate r_{max} (Slade and others, 1998; Millar and Meyer, 2000; Niel and Lebreton, 2005; Gedamke and others, 2007), and the assessment could use any of these depending on the data available for a species. For example, if a bird species has temporal data showing its recovery after reaching a small population size, the methods in Gedamke and others (2007) or Millar and Meyer (2000) might be useful. Runge and others (2009) used the method of Slade and others (1998) to estimate r_{max} for black vultures (*Coragyps atratus*). Runge and others (2009) assigned probability distributions to input variables and used Monte Carlo simulations to propagate that uncertainty when estimating $i_{,ax}$. Ultimately, the final decision on the approach used to estimate r_{max} for a species will be a key undertaking of the assessor and assessment panel. For marine mammals, GAMMS recommends using default values previously developed for groups of species (for example, pinnipeds or cetaceans) in the absence of other compelling data. It may be possible to develop default values for groups of birds and bats on the basis of body size and other life history parameters, but the authors of this report are not aware of such values being estimated.

As with r_{max} , N_{min} can be calculated by using a variety of approaches that are often dependent upon the data available. In the Marine Mammal Protection Act (16 U.S.C. § 1362(27), as amended in 2004), N_{min} is defined as follows:

> (27) The term "minimum population estimate" means an estimate of the number of animals in a stock that—
>
> (A) is based on the best available scientific information on abundance, incorporating the precision and variability associated with such information; and,
>
> (B) provides reasonable assurance that the stock size is equal to or greater than the estimate.

Taylor (1993) and Wade (1998) used simulations to study how setting the lower limit of population size (on the basis of a percentile from the distribution around an average abundance) affected how a PBR could be used to meet conservation criteria. Wade (1998) showed that setting N_{min} to the 20th percentile (the lower bound of a log-normal 60-percent confidence limit), resulted in 95 percent of simulation runs remaining at or above the target population size and the recovery of populations in simulations where population sizes were started below the target population size. The target population size was defined as the population size that resulted in the "optimum sustainable population," which is defined in the Marine Mammal Protection Act as—

> (9) ... the number of animals which will result in the maximum productivity of the population or the species, keeping in mind the carrying capacity of the habitat and the health of the ecosystem of which they form a constituent element.

The 20th percentile approach is now recommended by GAMMS and is formalized into a specific equation for calculating N_{min} :

$$N_{min} = N/e^{\left(0.842\left(ln(1+CV(N)^2)\right)^{1/2}\right)} \quad (5)$$

where N is an unbiased abundance estimate and $CV(N)$ is the coefficient of variation of the abundance estimate. When possible, the assessors will use this method when estimating N_{min}.

The term F was originally described as a recovery factor (Wade, 1998), but it more broadly represents management goals and has biological implications (Runge and others, 2009). For example, values of F between 0 and 2 result in PBR values that, if met, will achieve a sustainable population. The resulting sustainable population size will vary with F. Assuming linear density dependence of the population growth rate, at $F = 0$ and 1, the population will equilibrate at carrying capacity (K) and K 2, respectively. As F increases towards 2, fatalities approach r_{max} , and at $F = 2$, the PBR is $r_{max} \times N_{min}$ and theoretically the population size should decline to 0. Thus, values of F near 2, but less than 2, would result in very small equilibrium population sizes, well below the population's carrying capacity.

The relation between F and the equilibrium population size can be used to set F for particular policy objectives. Under the MMPA, F is restricted to be between 0.1 and 1. This range in F is based primarily on extensive simulation studies of marine mammal populations (Barlow and others, 1995; Wade, 1998) and the management objectives defined in the MMPA. The results of the GAMMS simulation analyses suggest a value for F of 0.1 for stocks considered endangered; 0.5 for stocks considered depleted, threatened, or unknown; and 1.0 for stocks known to be at their target population size, stocks that are increasing, or stocks that are not decreasing and are harvested only by subsistence hunting.

This methodology will follow the GAMMS suggestions for *F* with modifications for birds and bats. Endangered species are assigned a value for *F* of 0.1 to reflect a management objective of recovery. Threatened species, species showing population declines, or species with an unknown population status are assigned a value of 0.5. Species that are not listed, exhibit stable or growing populations, or are not considered overabundant are assigned a value of 1. Finally, overabundant species with management goals directed towards reducing population size are assigned a value of 1.5.

2.4.2. Implementation

To assess the risk of population decline for a species, the assessment will calculate the risk ratio (RR), which is simply the annual estimated fatalities divided by the PBR (Figure 6).

The risk ratio is the proportion of the PBR accounted for by fatalities at wind energy facilities. As this value increases from 0, the risk goes up; at a value of 1, fatalities are equal to the PBR. When the RR is greater than 1, fatalities are greater than the PBR, and the population is expected to decline. Uncertainty in the RR will be estimated by sampling across the uncertainty in the PBR and the uncertainty in estimated fatality to calculate a mean RR and an accompanying confidence interval.

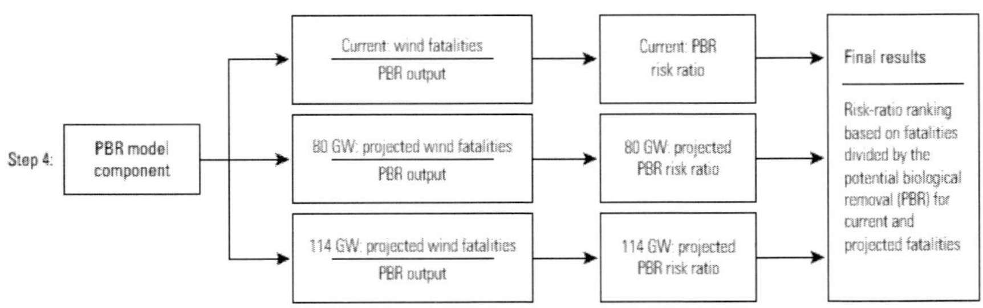

Figure 6. Flowchart of the PBR model component of the assessment methodology. GW, gigawatts; PBR, potential biological removal.

As with the FT metric in the prioritization component (section 2.2.3.1), the observed number of fatalities is currently calculated by using an approach that does not correct for differences in species detection rates and can cause significant biases in the estimated number of fatalities. The use of these current estimates could lead to unreliable assessment results at the national or regional level.

The methodology includes the RR to allow the identification of species at different levels of risk. However, the broader implications of the RR and PBR for the overall status of a species are critical to consider. A species with a low RR may still be imperiled because fatalities from other sources push the species above the PBR, or because other processes not measured by the PBR, such as habitat loss, are affecting a species. Thus, the RR measures only the impact of fatalities from wind turbines in the United States on a species, not the effects of other sources of fatality or other effects of wind turbines.

During an implemented assessment, a number of activities would occur to better improve estimates of the PBR. First, assessors would meet with members of regulatory agencies to better define management objectives and possibly revise the default values of F. Second, as r_{max} is estimated for more species and the range of r_{max} values is better understood, simulation studies similar to those performed during development of the GAMMS recommendations could be performed to determine values of F and N_{min} that would meet established management objectives. Third, a meta-analysis of r_{max} estimates could be conducted to perhaps estimate defendable default values of i_{max} by taxon or life-history type. Fourth, the nature and shape of density dependence in birds and bats would require study. Most PBR applications assume linear density dependence, and the methodology currently follows this assumption. Additional research on birds and bats could help verify or refine this assumption.

3.0. Future Projections

The methodology uses projections of newly installed capacity of wind energy facilities to extrapolate the impacts of future wind development on wildlife. Both the demographic model and the risk ratio are rerun with the updated fatality estimates based on the projections. A key assumption is that fatalities will increase as additional turbines are installed. This may not be true if new facilities can be developed in a manner that avoids and minimizes impacts to wildlife.

A number of organizations have projected future levels of installed wind capacity for the United States. These projections produce a single estimate of new capacity for the entire Nation. An exception is a study by the National

Renewable Energy Laboratory (NREL; Hand and others, 2012), which estimated new wind capacity for each State under different levels of national renewable energy production. The NREL researchers investigated varying levels of new capacity and used an economic-demand-based model to predict installed capacity of wind energy by State through time.

Projecting future levels of annual fatalities from wind energy development involves taking existing fatality rates (deaths per megawatt per year) and adjusting them on the basis of predicted levels of installed wind turbine capacity in megawatts (MW). The most basic approach would simply multiply existing fatalities by the percent increase in installed capacity (for instance, if 100 animals die per year and installed capacity will double, then 200 animals are projected to die per year).

The approach in this methodology differs between birds and bats because there is more statistical modeling of bird fatality than bat fatality. Estimates of overall bat fatality exist and suggest that fatalities may be influenced by a number of factors (Baerwald and Barclay, 2009; Arnett and Baerwald, 2013; Hayes, 2013; Smallwood, 2013; Huso and Dalthorp, 2014). However, statistical models of these factors do not exist and cannot be used for projecting. Bat fatality data are being compiled to allow such modeling (see appendix 1), and these models may be available by the time an assessment is implemented. For now, only raw extrapolations based on national increases in wind energy production can be performed for bats, and extrapolations will be referred to from here on For birds, the approach uses an existing statistical model of fatalities (Loss and others, 2013). The model requires an estimate of the number of turbines and their heights in each of four regions of the contiguous United States (California, East, Great Plains, and West excluding California, as in Figure 1 of Loss and others, 2013). To calculate these estimates, the approach first partitioned the projected new capacity among the regions and added this to the existing capacity in each region. Then each regional capacity was converted into a prediction of the number of turbines by dividing the regional capacity by an estimate of the average size (in megawatts of capacity) of each turbine. Finally, because turbine height is correlated with turbine capacity, this relation was used to predict the average turbine height in each region. These values were then used in the statistical model to project future fatalities.

Thirty-five projections of future wind energy capacity in the United States in 2025 were chosen to estimate medium and high build-out scenarios, which are estimates of the amount of new wind energy facilities installed in the United States. Of the 35 projections, 3 were developed by private companies, 2 by the

International Energy Agency (IEA), and 30 by the U.S. Energy Information Administration (EIA); details for the projections were provided by the U.S. Energy Information Administration (2014a) and the International Energy Agency (2012). The build-out scenarios developed by the EIA included factors such as energy demand, regulatory limits on greenhouse-gas emissions, and changes in the amount of electricity produced by nuclear power. The medium scenario was defined as the average installed capacity across all 35 projections (80 GW), whereas the high scenario represented the value at the 95th percentile of the distribution of projections (114 GW). The projected installed capacities (80 and 114 GW) were national estimates that were partitioned across the regions of the United States (California, East, Great Plains, and West excluding California) used in the statistical models developed by Loss and others (2013). To do so, the projected additional national capacities (80 − 61 GW = 19 GW or 114 − 61 GW = 53 GW) were multiplied by the predicted proportions of capacity in each region in the Loss and others (2013) models and then added to the existing installed capacity in each region (Figure 7).

The proportional estimates came from predictions of installed capacity in each State from 33 simulations of wind energy growth modeled by the NREL (Hand and others, 2012). Predictions for each State were summed within a region, then the proportion of the total installed capacity was estimated for each region. These proportions were then averaged across the 33 simulations.

Once national projections of additional installed capacity were allocated to each region, the regional capacity values (in gigawatts) were converted to the number of turbines expected to be installed in each region. Current information on the average capacity of turbines (in megawatts) and its trend through time was used to convert regional capacity (in gigawatts) into the number of turbines. A linear increase in average capacity of turbines was assumed on the basis of the rate of increase in average capacity of active turbines installed in each region since 2005 (Figure 8). The projected future regional capacity was then divided by the estimated capacity per turbine to determine the number of installed turbines in each region.

Because increasing turbine height correlates with increases in fatalities (Loss and others, 2013), we also modeled the projected height of future turbines. Turbine height is related to turbine capacity (Figure 9). A power function was fitted to the heights and capacities of turbines that were active in each region in 2014. This relation was used to estimate average turbine height based on average capacity (in megawatts) in each scenario.

The predicted number of turbines in each region and their associated heights for each wind energy projection scenario were input into the statistical model developed by Loss and others (2013) to estimate the number of annual avian fatalities expected at onshore windpower facilities in the United States in 2025 (Table 2). Although it would be possible to develop more sophisticated methods relying on distributions of turbine capacities and heights, assumptions about the shapes of these distributions would be speculative. Because estimates produced using mean heights and capacities for current turbines differed by less than 3 percent from those produced using numbers of turbines of different heights and capacities, the added detail is unlikely to greatly improve the estimates of fatalities.

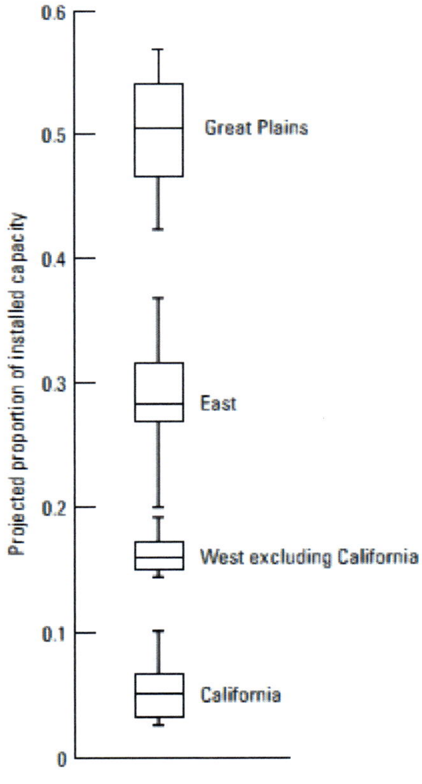

Figure 7. Boxplot showing projected proportions of wind energy capacity based on 33 projections for 2025 by the National Renewable Energy Laboratory (Hand and others, 2012), partitioned by U.S. region (from Loss and others, 2013). The box contains the middle 50 percent of values, with a line at the median. Whiskers mark the maximum and minimum values. Average values were used when projecting future wind capacity in each region.

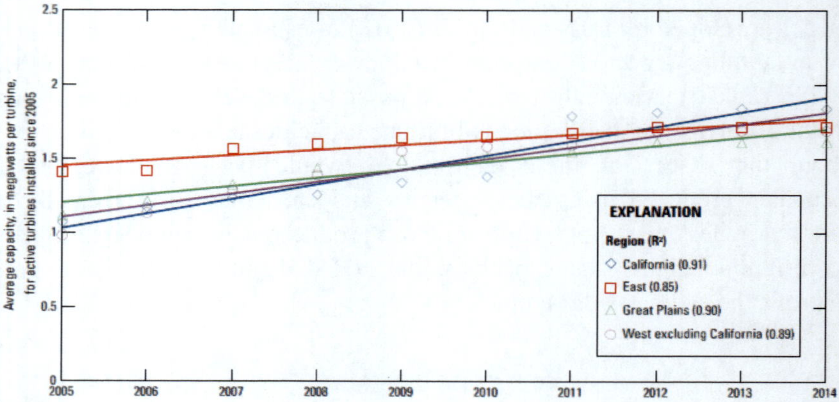

Figure 8. Graph showing the average turbine capacity of active turbines installed since 2005 in the four regions of Loss and others (2013) in the contiguous United States. The lines represent a linear regression run for each region. R2 is the coefficient of determination and indicates how well a model fits the data.

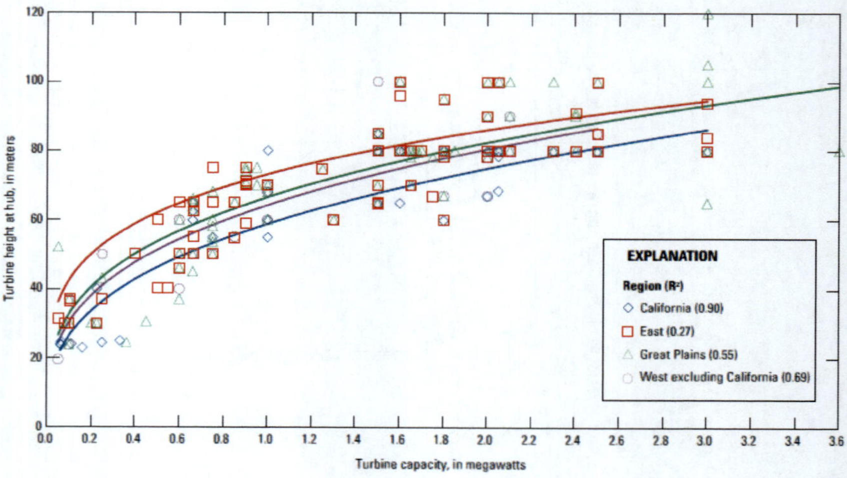

Figure 9. Graph showing the relation between turbine height (hub height) and capacity in the four regions of Loss and others (2013) in the contiguous United States for turbines active in 2014. Curved lines represent power functions (linear regression on a log-log scale) relating turbine capacity to hub height in each region. R^2 is the coefficient of determination and indicates how well a model fits the data.

Table 2. Projected number of annual avian fatalities expected at onshore windpower facilities in the United States in 2014 and in 2025

[GW, gigawatt; max, maximum; min, minimum; NA, not applicable]

Year installed-capacity scenario)		Mean annual avian fatalities (min–max)		Increase in fatalities (percent)
2014	(62.3-GW capacity)	182,000	(123,000–240,000)	NA
2025	(80-GW capacity)	233,000	(115,000–323,000)	28.0
2025	(114-GW capacity)	333,000	(206,000–460,000)	83.0

4.0. Overall Methodology Development and Validation

As noted in section 1.3, the methodology has a number of elements that should be checked during the implementation of an assessment. Furthermore, specific elements of the methodology will need to be refined if an assessment is implemented. These refinements include guidelines for (1) estimating demographic parameters when data are sparse, (2) developing species distribution maps by a common method, (3) estimating species-specific fatalities, and (4) selecting species after prioritization.

While developing the methodology, specific approaches were implemented for the case study species described below, but some of the approaches used may not be optimal as data on more species are used in the methodology. For example, the components of the methodology produce consistent results across the case study of six species, but as more species are prioritized and their data are used in the demographic model and risk ratio components, consistency across the components should be checked. In addition, the results of the demographic and PBR components could be compared with the outcomes of more detailed research as they become available. Customized population models are being developed for several species that would also be assessed with this methodology, including the golden eagle and whooping crane (Butler and others, 2013), and these models could be used to check that the simpler models used in the methodology produced qualitatively consistent results. If consistency declines, modifications may be necessary, depending on the reasons.

The methodology would also be much improved, and would produce more reliable results, if four key issues could be resolved. First, robust, species-specific fatality estimates do not currently exist. The current approach used to estimate species-specific fatality rates (section 2.3) may produce unknown biases because it does not account for species detect-ability. Second, population-size estimates are poor for most species, particularly bats. The uncertainty around the turbine mortality rate could be decreased considerably with better population-size information. Third, range maps describing a species' relative abundance across space and through time are not currently available for most species. Because these maps would be used to compare species, a consistent method to produce them on the basis of widely available location data (for example, from eBird; Cornell Lab of Ornithology, 2015) would be an invaluable contribution to the assessment. Fourth, the methodology would be improved if scientists could develop a greater understanding of the role of additive versus compensatory mortality in birds and bats. The methodology currently uses the assumption that all of the fatalities from wind energy facilities are additive. This assumption is likely violated for many species, but assuming that fatalities from wind energy facilities are compensatory also essentially means that one assumes that the fatalities from wind energy facilities have no effect on a species. Instead, the methodology evokes a precautionary assumption, and this should be considered when interpreting the results of the methodology.

Finally, during an official assessment, the methodology would require input from decision makers at three points in the process. These inputs are unavoidable and required for those steps in the methodology that are primarily policy, not scientific, decisions. First, the ranking systems in Table 1 are subjective and currently based on the authors' perception of a need to consider the effects of wind energy development ahead of conservation status. This ranking system requires external review and adjustment prior to an assessment. Second, a decision must be made about what species should be further analyzed after prioritization; that is, decide how far down the list of average ranks of relative risk the assessment should proceed because species below the designated stopping point will not be included in the demographic model or risk-ratio analyses. Third, the methodology currently uses the guidelines associated with marine mammals as default values for F when calculating the PBR. The parameter F is associated with levels of acceptable risk and should be set by regulators, not the USGS. Thus, the USGS would require input from regulatory agencies to determine whether the existing values of F are appropriate.

5.0. Test Case Results

The results described here should not be considered definitive statements of risk associated with wind energy development. The input data were gleaned from a variety of sources and not thoroughly vetted as they would be during a formal assessment. Because of this, species names are not reported. The results presented here give readers an example of the output an assessment would produce, but they may not correctly characterize the impact of wind energy development.

5.1. Introduction

Data for six avian species were processed through the complete assessment methodology to demonstrate the required input parameters and the methodology outputs. These species were selected because data were readily available and because analysis of these species may produce a diverse range of potential results in the assessment. The following section presents test case results from the prioritization method, the demographic model, and the risk ratio. The results are shown for three scenarios defined by installed capacity as follows: (1) current (2014) wind energy development (62.3 GW), (2) moderate wind energy development for the year 2025 (80 GW), and (3) high wind energy development for the year 2025 (114 GW). See section 3 for details on the methods used to develop the scenarios.

5.2. Data Sources

Published research, data from resource-management agencies, and reference works were used to develop the parameter values. Conservation-status information was found at Web sites and in reports of the IUCN (International Union for Conservation of Nature, 2014), the USFWS (U.S. Fish and Wildlife Service, 2014), and individual State wildlife action plans (Association of Fish and Wildlife Agencies, 2015). Population size (*N*) for all species came from the Partners in Flight Science Committee (2013), and 30 percent of the population size estimate was used as a rough approximation of N_{min}. Either direct estimates or data used to calculate demographic parameters (survival, maternity, r_{max}, and age at first reproduction) were found in Poole

(2005) and Michel and others (2006). When possible, maternity (*m*) was calculated as the product of nest success, clutch size, hatchability, clutches per year, and presumed equal sex ratios at hatching. If data on hatchability or the number of clutches were lacking, the parameters for them were set to 0.9 and 1, respectively. The age of first reproduction (*a*) for these species was presumed to be the first breeding season after fledging unless evidence suggesting otherwise was available.

The percentage of a species' range that overlaps with the locations of wind turbines (*p*) was calculated as described in section 2.4, using North American Breeding Bird Survey relative abundance maps (Sauer and others, 2015) and the USGS wind turbine location data (Diffendorfer and others, 2014).

The numbers of fatalities (*n*) were estimated from turbine mortality data compiled from multiple sources by Loss and others (2013) following the approach described in section 2.2.3.1.

5.3. Prioritization Results

Input data, calculated conservation status, and turbine-risk metrics for the prioritization component are listed in Tables 3 and 4. Although the data sources described here may be the same as those used in an actual assessment, a vetting process by an assessment panel would ensure that the data were appropriately selected and used. The vetting process was not included in this test case.

Of the six species investigated by using the prioritization method, three had average direct-risk ranks of less than 5, and two had average indirect-risk ranks of less than 5 (Table 4). Raptor 1 ranked the highest of all species for average direct risk because it had the highest FT and FRI. The corvid and raptor 2 ranked high on direct risk because of their high FRI values. Both songbirds ranked low on direct risk despite an elevated conservation status because few die as the result of collisions with turbines, and their "fast" life histories resulted in low FRI values. Species that use fewer habitats and that have conservation status (both songbirds) scored higher on the average indirect risk. The three demonstration species with direct-risk ranks of less than 5 were evaluated by using the demographic model and potential biological removal.

5.4. Demographic Model Results

Following the high-priority ranking from the prioritization step, the corvid and both raptors were analyzed with the demographic model. The input parameters that could be used for demographic modeling are provided in Table 5. The analysis was conducted following the process described in section 2.3.

The key results of the demographic model are presented in Table 6. Because all three species had adequate time series to calculate an observed trend, the matrix model was not used. Note that the trend was calculated by using data up to 1990, prior to large numbers of turbines being installed, but this endpoint may not be appropriate in an actual assessment given potential changes in population status since 1990 due to factors other than wind energy development. The change in the population trend and the resulting projected trend with additional mortality from wind energy development are reported as means with 90-percent confidence intervals for each of the three scenarios. Finally, the increase in the percentage of population growth rate values that were less than 1 is reported along with the percentage of population growth rate values less than 1 with wind-facility-related mortalities for each scenario.

The corvid showed very small responses to fatalities from wind turbines, even in the wind-energy-development scenarios involving a high installed capacity. The corvid ranked highly during prioritization because its life-history parameters produced a high FRI. However, relative to its population size, a small fraction of individuals were projected to be killed (Table 5), resulting in small estimates of added mortality and minimal changes in population trend. The projected declines in population trend for the corvid ranged from 0.00002 to 0.0003 across current and projected levels of wind energy generation (Table 6).

Both raptors had long-term declines in population size with observed mean population trends less than 1. Current levels of wind energy facilities had a larger negative effect on raptor 1 (mean decline=0.016) than raptor 2 (mean decline=0.004). For raptor 1, with current levels of wind energy facilities, the mean observed population trend was projected to decrease from 0.989 to 0.973, while raptor 2 declined from a mean of 0.995 to 0.991. As expected, declines in population trend increased with higher levels of wind energy development. For raptor 1, the model also projected large increases in the percentage of population growth rates less than 1 between the trends without and with wind energy development.

Table 3. Input data for calculated conservation status and direct- and indirect-risk metrics for the prioritization component of the test case

[Species are anonymized to reduce the chance of inferences being developed prior to full vetting of the input data and the methodology. No., number; SGCN, species of greatest conservation need]

Species	No. of States that consider species as an SGCN	No. of States in which species occurs	No. of turbine collision fatalities for species, annually (n)	Population size of species in United States, in millions (N)	Adult survival (s)	Percentage of species range overlapping with locations of wind turbines (p)	Maternity (m)	Age at first reproduction (a)	No. of habitats (h)
Corvid	2	49	260	17	0.865	4.04	0.60	2	7
Raptor 1	14	49	7,594	1.7	0.490	5.42	1.29	1	10
Raptor 2	34	50	519	0.5	0.700	4.47	1.02	1	8
Raptor 3	1	49	4,933	5	0.770	4.80	0.51	1	6
Songbird 1	24	48	130	0.8	0.529	3.74	1.33	1	2
Songbird 2	29	46	1,363	11	0.610	2.89	0.99	1	3

Table 4. Output values of conservation status and direct- and indirect-risk metrics and the resulting average rank for risk from collision fatalities and risk from habitat change (lower values indicate higher priority) from the prioritization approach

[IUCN, International Union for Conservation of Nature; NA, not applicable; percent SGCN, percentage of States where species is present that consider it a species of greatest conservation need (SGCN); USFWS, U.S. Fish and Wildlife Service]

Species	Conservation status			Percentage of fatalities in United States caused by turbines (FT)	Fatality-risk index (FRI)	Indirect-risk index (IRI)	Average rank for direct risk	Average rank for indirect risk
	Percent SGCN	USFWS designation	IUCN designation					
Corvid	4.08	NA	Least concern	0.0113	13.518	0.578	3.84	7.54
Raptor 1	28.57	NA	Least concern	0.8751	10.627	0.542	2.93	6.87
Raptor 2	68.00	NA	Least concern	0.3462	4.393	0.558	4.14	5.48
Raptor 3	2.04	NA	Least concern	0.4280	8.334	0.800	5.41	7.04
Songbird 1	50.00	Bird of conservation concern	Least concern	0.0345	2.173	1.869	6.72	2.25
Songbird 2	63.04	Bird of conservation concern	Least concern	0.0318	2.934	0.964	5.99	4.45

Table 5. Parameter inputs for the matrix model for three prioritized species in the test case investigation

[Nest success is the proportion of nests that hatch at least 1 egg. Hatchability is the proportion of eggs that hatch in successful nests. No., number; NA, not available]

Species	No. of turbine fatalities	Population size of species in United States	Juvenile survival (s_j)	Adult survival (s_a)	Turbine mortality rate (c)	Age at first reproduction (years)	Clutch size	Nest success	Hatchability	Clutches per year	Maternity (m)
Corvid	260	1.7×10^7	0.63	0.86	0.002	2	4	0.33	0.91	1	0.60
Raptor 1	7,594	1.7×10^5	0.34	0.49	0.447	1	4.6	NA	NA	1	1.29
Raptor 2	519	5.0×10^5	0.41	0.70	0.104	1	4.4	0.51	NA	1	1.02

Table 6. Results of the demographic model for 2014 and projected levels of wind development

[The observed population trend, projected decrease in population trend, and expected population trend are reported with 90-percent confidence intervals. Installed capacity in gigawatts (GW) for each scenario is as follows: current (2014) scenario, 62.3 GW; medium-capacity scenario for 2025, 80 GW; high-capacity scenario for 2025, 114 GW. w/o, without; λ, population growth rate; <, less than]

Observed population trend w/o wind energy development, 1966–1990	Scenario	Projected decrease in population trend	Expected population trend with wind energy development	Projected increase in percentage $\lambda<1$	Expected percentage $\lambda<1$
		Corvid			
1.005 (0.983–1.027)	Current	7×10^{-5} (2×10^{-5}–2×10^{-4})	1.005 (0.983–1.026)	0.2	35.1
	Medium	1×10^{-4} (2×10^{-5}–2×10^{-4})	1.005 (0.983–1.026)	0.3	35.2
	High	1×10^{-4} (3×10^{-5}–3×10^{-4})	1.005 (0.984–1.026)	0.4	35.3
		Raptor 1			
0.989 (0.950–1.028)	Current	0.016 (0.003–0.035)	0.973 (0.931–1.014)	18.2	85.7
	Medium	0.022 (0.004–0.047)	0.967 (0.923–1.011)	21.1	88.6
	High	0.031 (0.007–0.067)	0.958 (0.907–1.006)	25.0	92.5
		Raptor 2			
0.995 (0.912–1.077)	Current	0.004 (0.001–0.008)	0.991 (0.909–1.073)	3.1	56.9
	Medium	0.005 (0.001–0.012)	0.990 (0.907–1.072)	4.1	57.9
	High	0.007 (0.002–0.016)	0.988 (0.905–1.070)	6.0	59.8

5.5. Potential Biological Removal and Risk Ratio

As with the demographic model, the three high-ranking species from prioritization were analyzed by using potential biological removal (PBR) and risk ratios. The three species showed considerable differences in both the PBR and the risk ratio (Table 7).

The corvid had the largest PBR across all values of *F*, followed by raptor 1, then raptor 2 (Figure 10). To calculate risk ratios, an *F* value of 0.5 was selected because all three species showed declines or uncertain trends in the last 11 years of available BBS data (2002 through 2012; Sauer and others, 2015).

With *F* at 0.5, raptor 1 had a high risk ratio and the confidence interval overlapped 1 at current levels of wind energy development (Figure 11). Under both projections of future wind energy development, the best estimates of the risk ratio were at or over 1. Raptor 2's risk ratio at current levels of wind energy development was moderate, and the upper confidence interval almost overlapped 1. The risk ratio increased with high levels of projected wind energy development. Finally, the corvid risk ratio was near 0, indicating low risk for this species.

This demonstration of the methodology indicates that of the six species originally considered, three were prioritized for a more in-depth investigation. Of those three, raptor 1 indicated a higher risk than the other species both in potential declines in population trend and in the risk ratio. Raptor 2 indicated moderate levels of potential decreases in trend and a moderate to high risk ratio. The corvid was prioritized on the basis of its high FRI score, but because its fatalities were projected to be low relative to its population size, it showed negligible effects from wind turbines.

5.6. Implications of the Test Case

The USGS approach outlined here provides a quantifiable and replicable means of determining the relative risk and subsequent impact of adverse effects of wind energy facilities on volant wildlife. Output from the approach (risk indexes and estimates of the change in population trend for current and future projected wind energy development) could be provided for all high-priority bird and bat species for which sufficient data have been gathered. This approach relies on the development of new sources of information, such as species-specific fatality data and a suite of basic population parameters for the less studied bat species affected by wind energy generation. This approach should be used iteratively, updating risk as new information regarding species exposure and wind energy capacity changes. An example of a potential summary table that includes the major final outputs for the test case species is presented in Table 8.

Table 7. Parameter inputs and outputs for potential biological removal and risk ratios at current levels of wind energy development for three prioritized species

[Values in parentheses represent the 95-percent confidence interval. *F*, recovery factor; N_{min}, minimum population size; PBR, potential biological removal; r_{max}, maximum annual population growth rate; risk ratio, turbine fatalities divided by the PBR]

Species	No. of Turbine fatalities	*F*	r_{max}	N_{min}	PBR	Risk ratio
Corvid	260	0.5	0.07	(0.06–0.15)	5,100,000 89,250 (73,950–91,250)	0.003 (0.001–0.004)
Raptor 1	7,594	0.5	0.08	(0.03–0.10)	510,000 10,124 (3,824–12,750)	0.750 (0.463–2.055)
Raptor 2	519	0.5	0.030	(0.014–0.06)	150,000 1,106 (525–2,250)	0.469 (0.180–0.988)

CONCLUSION

The USGS Wind Energy Impacts Assessment Methodology (WEIAM) project team created an assessment methodology to provide both qualitative and quantitative metrics related to the effects on birds and bats from wind energy development at the national scale and species level. This work built on a variety of quantitative approaches focused on ecological risk assessment to provide a regional to national perspective that will inform decision makers, industry, and the public.

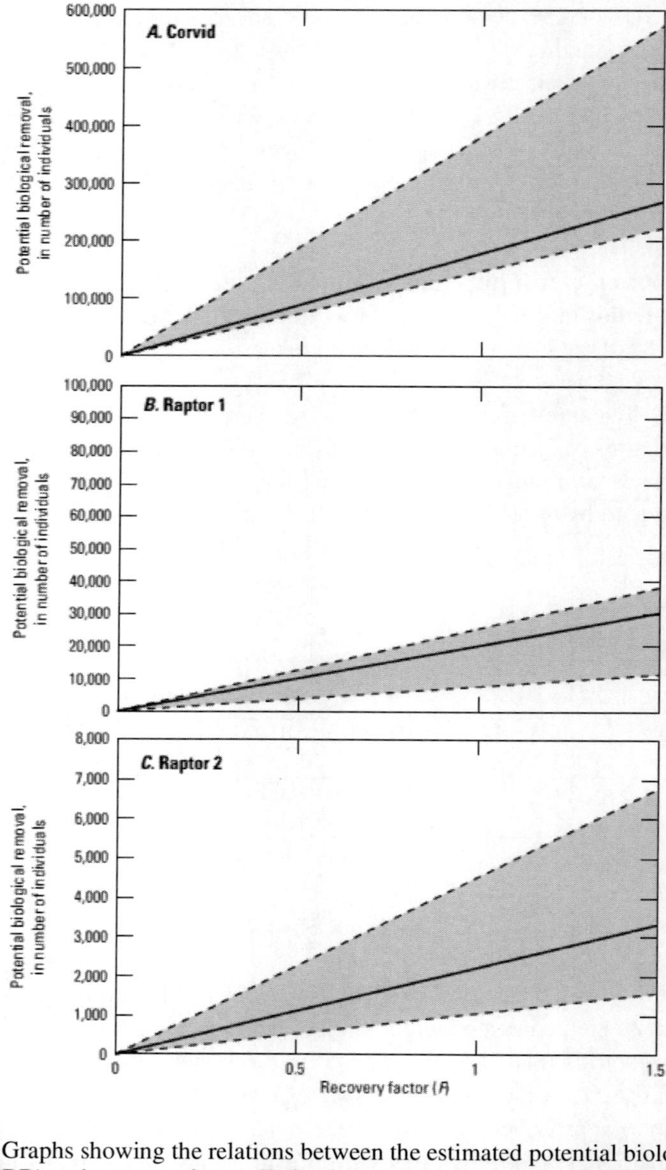

Figure 10. Graphs showing the relations between the estimated potential biological removal (PBR) and recovery factor (F) for three species at current (2014) levels of wind energy development. The solid black line represents the best estimate of the PBR; dashed lines represent the lower and upper confidence limits. A, Corvid; B, Raptor 1; C, Raptor 2.

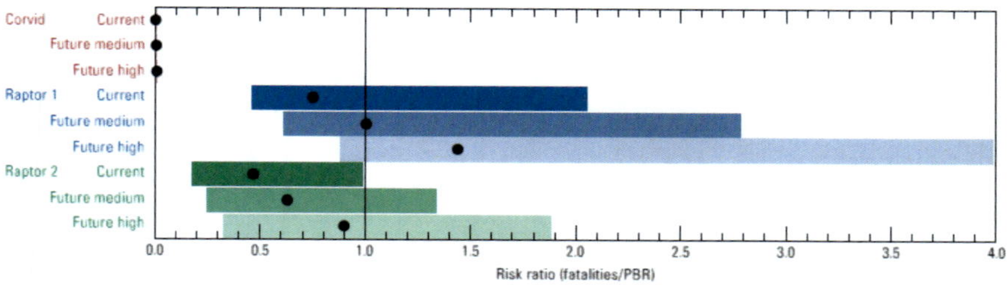

Figure 11. Graph showing the ranges of relative risk ratios when the recovery factor (F) equals 0.5 for three species at current (2014) levels of installed capacity and for both medium- and high-capacity scenarios for 2025. The black circles represent the best estimate, whereas the colored bars represent the range spanned by the upper and lower confidence intervals. The projected scenarios are as follows, by installed capacity in gigawatts (GW): current (2014), 62.3 GW; future medium (2025), 80 GW; future high (2025), 114 GW. PBR, potential biological removal.

Table 8. Summary table of the main output values for all components of the methodology: prioritization, the demographic model, and potential biological removal

[Installed capacity in gigawatts (GW) for each scenario is as follows: current (2014) scenario, 62.3 GW; medium-capacity scenario for 2025, 80 GW; high-capacity scenario for 2025, 114 GW. Risk-ratio values are presented as an average, and values ratios greater than 1 indicate a decrease in population size. NA, not applicable; PBR, potential biological removal; λ, population growth rate; --, species were not highly prioritized and, therefore, no quantitative model outputs are available]in parentheses represent the 95-percent confidence interval; risk

		Species prioritization component		Demographic model component	PBR model component
Species	Scenario	Average rank for direct risk	Average rank for indirect risk	Projected increase in percentage λ<1	Risk ratio
Corvid	Current	3.84	7.54	0.2	0.003 (0.001–0.004)
	Medium	NA	NA	0.3	0.004 (0.001–0.005)
	High	NA	NA	0.4	0.006 (0.002–0.007)
Raptor 1	Current	2.93	6.87	18.2	0.750 (0.463–2.055)
	Medium	NA	NA	21.1	1.006 (0.611–2.780)

Table 8. (Continued)

Species	Scenario	Species prioritization component		Demographic model component	PBR model component
		Average rank for direct risk	Average rank for indirect risk	Projected increase in percentage $\lambda<1$	Risk ratio
	High	NA	NA	25.0	1.438 (0.882–3.983)
Raptor 2	Current	4.14	5.48	3.1	0.469 (0.180–0.988)
	Medium	NA	NA	4.1	0.629 (0.248–1.333)
	High	NA	NA	6.0	0.900 (0.325–1.884)
Raptor 3	Current	5.41	7.04	--	--
Songbird 1	Current	6.72	2.25	--	--
Songbird 2	Current	5.99	4.45	--	--

The assessment methodology described in the preceding sections is an initial attempt to quantify the significance of the effects of wind energy development on species population trends. Uncertainty is captured in the input ranges for the model parameters and reflected in the probabilistic assessment output. The test case provided is only an example of the method and is not an official assessment result. As the science of impact research matures, this method may be updated and improved to reflect new knowledge in this rapidly growing field of study. A national assessment of impacts to species is not currently part of this project; however, the creation of a methodology and process for assessment is a fundamental first step in any USGS assessment project.

The model described here and the resulting output should be considered only for a scientific assessment, not for a method to develop management strategies related to wind energy development for a particular species. The model produces an estimate, with uncertainty, of the effects on population trend or a risk ratio from observed fatalities at wind turbines. Decision makers prioritizing species for avoidance, minimization, and mitigation actions might use this information, and they might also include information about the feasibility, probability of success, and cost.

This approach should not replace detailed, species-specific studies or population models of those species garnering high levels of attention. Furthermore, the approach is not designed to estimate the total capacity levels of wind energy that could be installed across the Nation before species show population declines. Instead, the approach could help inform decisions related to—

1. Identification of those species at low risk from wind energy development.
2. Identification of those species that may be at risk from wind energy development.
3. Quantification of the expected decline in population trend for identified high-risk species from current and future levels of wind energy development.

ACKNOWLEDGMENTS

The authors thank all their fellow members of the Wind Energy Impacts Assessment Methodology project team: Maureen Thompson (Oklahoma State University) and Erik Paquin, Mark Bronzert, Jolene Keen, Donna Norton, Louisa Kramer, Zach Ancona, and Alisa Coffin (all of the U.S. Geological Survey, USGS). Reviews from Victoria Bennett, of Texas Christian University, and from USGS colleagues Paul Cryan, David Houseknecht, Todd Katzner, and Michael Runge were instrumental in creating the preliminary methodology document. Many meetings with personnel from a number of Federal agencies, concerned nongovernmental organizations, and industry provided the context in which to frame the assessment unit and outputs for the method. Discussions with Laura Ellison (USGS) and Paul Cryan during the development of the methods clarified our thinking on bats.

GLOSSARY

additive mortality – Mortality (often from humans) that occurs in addition to natural mortality. Additive mortality adds to the number of deaths that would have occurred naturally.

adult survival (*s*) - Probability that an adult animal survives from one year to the next.

age at first reproduction (*a*) - Age at which females become reproductively mature.

barotrauma - Damage to body tissue caused by the difference in air pressure around the turbine blade.

capacity - The rated maximum amount of energy a turbine is capable of generating.

compensatory mortality - Mortality (often from humans) that somehow is compensated for by changes (reductions) in natural mortality. Compensatory mortality does not add to the number of deaths that would have occurred from natural causes.

current conservation status - The methodology uses conservation status information from State wildlife action plans, where some species are considered "species of greatest conservation need"; other sources of information on conservation status include (1) the Red List of the International Union for Conservation of Nature (2014), (2) listing by the Endangered Species Act of 1973, as amended (16 U.S.C. §1531 et seq.), or (3) the list, "Birds of Conservation Concern 2008" (U.S. Fish and Wildlife Service, 2008).

demographic model - A mathematical model used to generate quantitative and probabilistic estimates of impacts to species populations associated with the additional fatalities caused by collisions with wind turbines.

fatalities from wind turbines (*n*) - The estimated number of fatalities per year caused by collisions with wind turbines. To date, n has been estimated by multiplying an estimate of total fatalities across all species by the proportion of the observed fatalities accounted for by a species.

fatality-risk index - FRI = $p/(m\ a)$ where p is the percentage of a species' range that overlaps with the locations of wind turbines, m is maternity (number of female offspring per adult female per year), and a is the age at which females become reproductively mature.

habitats (*h*) - Number of habitats used by a species.

juvenile survival - Chance of juveniles surviving from one year to the next.

maternity (*m*) - Number of female offspring per adult female per year.

overlap with wind turbines (*p*) - The proportion of a species' range that overlaps with the locations of wind turbines.

population size (*N*) - The total population size of the species that moves through the United States at some point in its life cycle.

population trend - The increasing, decreasing, or stable pattern of the population of a species through time.

potential biological removal (PBR) - An estimate of the total fatalities that can occur before the population of a species will decline below a target population size.

range map - Geographic distribution of a given species.

species of greatest conservation need (SGCN) Species considered to have an elevated conservation status in State wildlife action plans.

species prioritization - Qualitative approach to rapidly screen a large number of species and rank them in terms of the relative risk of a population-level consequence from wind energy development.

State wildlife action plan (SWAP) - Proactive plan, developed as a comprehensive wildlife conservation strategy, that helps to conserve nongame wildlife and vital natural areas before they become too rare and costly to protect.

time series - A group of data collected sequentially, usually at fixed intervals of time.

volant species - Winged species capable of flying.

APPENDIX 1. ONGOING RESEARCH TO IMPROVE FUTURE METHODOLOGIES

Introduction

Members of the U.S. Geological Survey's Wind Energy Impacts Assessment Methodology (WEIAM) project team investigated a number of approaches designed either to directly estimate the effects of wind energy development on wildlife or to understand key information gaps related to the issue of wind energy development's effects on wildlife. In addition to those initial investigations, research is ongoing and may help to improve future iterations of this methodology, as summarized in the following sections.

Flight Intensity as a Tool in Wildlife Impact Research

The use of airspace by flying animals is the nexus between wind energy facilities and the fatalities they cause. If a wind energy facility is located where few animals fly, then few fatalities are likely to result, whereas if a facility is located where flight activity by animals is intense, then there may be a much higher risk of fatalities. Accordingly, knowledge of airspace use is fundamental for assessing risks of wind energy development to wildlife. Identifying areas where flight intensity is high and where wind energy development has occurred or is likely to occur will greatly facilitate risk assessment. In terms of the methodology, increased understanding of airspace use relative to wind turbines would allow a more accurate estimate of population growth rate (λ) in the species prioritization metrics and would increase the accuracy of forecasts by predicting the proportion of a species' range that overlaps with the locations of wind turbines (p).

This research focuses on assessing the intensity of flight activity by animals and uses new types of information developed over the last decade. For example, the use of radiotelemetry unleashed a flood of information about how animals actually move (Duerr and others, 2012; Katzner and others, 2012; Lanzone and others, 2012; Miller and others, 2014). Fixed and mobile radar facilities provide information about mass movements of birds (Diehl and Larkin, 2005), as well as bats and insects (Kunz and others, 2008). New technologies such as acoustic monitors provide information on airspace use by bats (Britzke and others, 2013) and migrating birds (Blumstein and others, 2011), stable-isotope analysis of animal tissues sheds light on migration pathways of many species (Zimmo and others, 2012), and photosensitive geolocators can record information on approximate latitude and longitude traversed by animals (Stutchbury and others, 2009). Genetic markers, thermal cameras, Internet tools such as eBird (Cornell Lab of Ornithology, 2015), and cooperative monitoring projects such as the Nocturnal Flight Call Activity Index ("oldbird" at Old Bird, Inc., 2015) also provide information about movements of birds.

This research is evaluating the potential of the tools described above and data they generate for determining the intensity of low-elevation flight of birds, bats, and other flying animals. Information derived from these tools could be used to facilitate the assessment of the effects of wind energy facilities on wildlife. For example, the direct- and indirect-risk metrics used in the species prioritization component of the assessment would be improved with more accurate estimates of the proportion of species' ranges that overlap with the locations of wind turbines.

To evaluate the capabilities and limitations of each technology, our approach uses a matrix formulation that identifies strengths and weaknesses of the various technologies. No consolidated effort has been made to assess airspace use by animals, although numerous studies have considered isolated aspects of it. The new paradigm of considering airspace as habitat will facilitate fresh thinking about the broader issue and help to improve the methodology.

Indiana Bat and Generalized Bat Population Model

The Indiana bat (*Myotis sodalis*) is listed as "endangered" under the Endangered Species Act (16 U.S.C. § 1531 et seq.), and the little brown bat (*Myotis lucifugus*) is under consideration for listing due to population losses from white-nose syndrome. Currently, a population model and a graphical user interface have been developed for use by U.S. Geological Survey (USGS) scientists (Thogmartin and others, 2012; Erickson, Thogmartin, and Szymanski, 2014). This model has served as the basis for a theoretical framework providing insight into the effects of wind energy facilities on migratory patterns and spatial dynamics of bats (Erickson, Thogmartin, Russell, and others, 2014). WEIAM project scientists are working to parameterize and apply this model to bat populations by using data from the U.S. Fish and Wildlife Service. In addition to the effects of wind energy facilities, the concurrent effects of white-nose syndrome are included in the model, which can be expanded to assess the effects of climate change and other stressors on migratory bat species.

The current assessment methodology uses a matrix model for both birds and bats. Developing a highly customized model for Indiana bats and then generalizing it to a broader number of bat species benefits the methodology in two ways. First, the methodology can be partially validated by comparing its results to those from the more complex model. Second, if a generalized demographic model can be created for bats, then the methodology could use two types of generalized models (one for bats and one for birds) that better match the life history and ecological differences between these two groups.

Investigating Wind Energy Development Effects with Breeding Bird Survey Information

Observations of bird fatality events due to collisions with wind turbines are well documented. However, these incidental fatality events may or may not scale up to increasing the risk of significant decline or extinction at the population level.

WEIAM project scientists are developing a method that uses data from the North American Breeding Bird Survey (BBS; Sauer and others, 2015) for detecting effects of wind energy development at the population level. This method could assess whether wind energy development has had a detectable effect on the population trends of monitored bird populations through time.

The method could also assess the likelihood of wind energy development's contributions to future imperilment of avian species in the United States. The method has the potential to place the monitoring data collected through the BBS observation routes spatially and temporally within the context of wind energy development as it has been documented in the USGS wind turbine map and database (Diffendorfer and others, 2014).

The analysis removes observation error from BBS abundance indexes to estimate underlying population trends and variability through linear multivariate autoregressive state-space (MARSS) models. Spatial and temporal covariates related to the location and timing of wind energy development can be incorporated into the models to estimate current effects on observed trends at multiple spatial scales.

Unlike the current methodology, this approach is an empirically based attempt to directly measure the effects of wind energy development on bird species at the population level. The approach can also be generalized to assess non-avian species when similar monitoring data are available. This approach represents a strong addition to the methodology that relies on population models and could be included as an additional component if the results warrant its use. In addition, the population parameter estimates from the MARSS models may improve the observed population trend estimates used in the demographic model components. These analyses can assist in the parameterization of detailed demographic models that can explore specific scenarios for wind energy development. In addition, this approach is useful for making short- and longterm projections of risk to bird populations at both the regional and national scale for the purposes of management and priority ranking.

Improved Bat Fatality Data

Though national estimates of bat fatalities exist (Arnett and others, 2008; Arnett and Baerwald, 2013; Hayes, 2013; Smallwood, 2013; Huso and Dalthorp, 2014), no systematic and national-scale analyses have been conducted to identify correlates of bat fatality rates or to use fatality correlates to estimate national fatality. In addition, despite substantial variation in study designs among studies conducted at different wind energy facilities and in different years, little research has investigated how varying the approaches to study design, data collection, and statistical estimation at local sites influences large-scale fatality estimates.

Project researchers have reviewed roughly 120 original studies that have investigated collision fatality at individual U.S. wind energy facilities. These data may be used to conduct the following research: (1) to estimate bat fatalities at U.S. wind energy facilities (a) to identify significant predictors of fatality-rate variation and (b) to use these predictors to estimate national bat fatality (for example, following general approaches in Loss and others, 2013) and (2) to assess how varying the study design and data collection approaches (in fatality surveys, scavenger removal trials, and searcher detection rate trials) influences national estimates of bird and bat fatality.

Results from such analyses could be used to enhance the methodology or its implemented assessment in two ways. First, the resulting statistical models could be used to estimate future fatality under scenarios of increased installations of wind turbines. Second, depending on the results of the analyses regarding bias, it may be possible to estimate species-specific rates of fatality for bats, which would allow more precise estimates from multiple components in the general demographic model component of the assessment methodology.

APPENDIX 2. ALTERNATIVE MODELING APPROACHES CONSIDERED

Stochastic Population Model

The U.S. Geological Survey (USGS) initially developed a stochastic matrix population model that included juvenile and adult survival, as well as age-specific maternity. The model was used to project population size forward through time. To model the effects of wind energy facilities, mortality was incrementally increased until the population trends met a quasi-extinction breakpoint (a percentage of runs showing a given percent decline in 10 years). This level of mortality was then compared to the observed fatality rates at turbines to assess risk. The model was developed with ranges of uncertainty around each parameter and then was run repeatedly, sampling across the parameter ranges in each model run.

The USGS elected not to use this approach for a number of reasons. First, given parameter uncertainty, a large proportion of runs either declined rapidly or increased exponentially. In both of these cases, added mortality from wind energy facilities had negligible effects on population trends so the approach was a fairly coarse tool for focusing on potentially small population-level consequences from wind energy facilities. Second, deciding upon the appropriate level of stochasticity to include for each species was extremely difficult. For many species, parameter uncertainty included both process and sampling error, which biased variance, and therefore risk, upward. Third, the approach required the setting of quasi-extinction breakpoints to determine how frequently a given turbine mortality rate met criteria of decline, as well as determining what proportion of runs needed to be declining before the turbine mortality rate was considered to cause population declines. Setting these arbitrary breakpoints, as noted above in section 2, adds an additional layer of complexity to the assessment.

Complex, Highly Individualized Models

If time and resources were available, the USGS could develop models tailored to each species of interest that would be based on the current state of knowledge of that species and the available data. These models might include complex spatial structure, simulate individuals across space and time, and use time series of population size to help estimate values for model input (Conroy and others, 1995). The Indiana bat model (appendix 1) served as a case study for this alternative. Although highly tailored models would enhance the assessment and could be run for species of high conservation concern, resource limitations prevent this from being practical for a larger number of species.

Empirically Based Population Viability Analyses

When spatially replicated time series of species abundance exist, a variety of approaches can use the information included in them to forecast future population dynamics. These approaches include multivariate autoregressive state-space (MARSS) models, corrupted stochastic exponential growth with Gaussian errors (CSEG) (Holmes and others, 2007), and other approaches. These approaches vary considerably in methodology, with some simply estimating a population trend and others estimating demographic rates associated with a trend. Overall, these approaches are valid and useful given the goals of the USGS Wind Energy Impacts Assessment Methodology (WEIAM) project. The USGS did not use them mainly because they require time series with an adequate signal-to-noise ratio to detect a trend, which would require highly precise estimates of abundance or relatively long time series. Though time series exist for many bird species in the North American Breeding Bird Survey (BBS; Sauer and others, 2015), nocturnal and migratory birds are not included and most bat species do not have time series, which limits the generality of these methods. Future iterations of the assessment methodology may allow for multiple assessment approaches that all produce the same output but use different algorithms. If so, then this class of models could be used for those species with time series of abundance data.

References Cited

American Ornithologists' Union, 2014, American Ornithologists' Union checklist of North and Middle American birds: American Ornithologists' Union Web site accessed October 31, 2014, at http://checklist.aou.org/.

American Wind Energy Association, 2014, U.S. wind industry third quarter market report: American Wind Energy Association Web page accessed November 5, 2014, at http://www. awea.org/3Q2014.

Andelman, S.J., Groves, Craig, and Regan, H.M., 2004, A review of protocols for selecting species at risk in the context of US Forest Service viability assessments: Acta Oecologica, v. 26, no. 2, p. 75–83, accessed December 9, 2013, at http://dx.doi.org/10.1016/j.actao.2004.04.005.

Arnett, E.B., and Baerwald, E.F., 2013, Impacts of wind energy development on bats; Implications for conservation, *in* Adams, R.A., and Pedersen, S.C., eds., Bat evolution, ecology, and conservation: New York, Springer, p. 435–456, accessed on January 15, 2015, at http://link.springer.com/chapter/10.1007/978-1-4614-7397-8_21.

Arnett, E.B., Brown, W.K., Erickson, W.P., Fiedler, J.K., Hamilton, B.L., Henry, T.H., Jain, Aaftab, Johnson, G.D., Kerns, Jessica, Koford, R.R., Nicholson, C.P., O'Connell, T.J., Piorkowski, M.D., and Tankersley, R.D., Jr., 2008, Patterns of bat fatalities at wind energy facilities in North America: Journal of Wildlife Management, v. 72, no. 1, p. 61–78, accessed May 14, 2013, at http://dx.doi.org/10.2193/2007-221.

Association of Fish and Wildlife Agencies, 2015, State wildlife action plans (SWAPs): Association of Fish and Wildlife Agencies Web site accessed February 20, 2015, at http:// www.teaming.com/state-wildlife-action-plans-swaps.

Athreya, K.B., and Karlin, Samuel, 1971, On branching processes with random environments. I. Extinction probabilities: Annals of Mathematical

Statistics, v. 42, no. 5, p. 1499–1520. [Also available at http://dx.doi. org/10.1214/ aoms/1177693150.]

Baerwald, E.F., and Barclay, R.M.R., 2009, Geographic variation in activity and fatality of migratory bats at wind energy facilities: Journal of Mammalogy, v. 90, no. 6, p. 1341– 1349, accessed April 17, 2013, at http://www.mammalogy.org/uploads/Baerwald%20and%20Barclay%202009.pdf.

Baerwald, E.F., D'Amours, G.H., Klug, B.J., and Barclay, R.M.R., 2008, Barotrauma is a significant cause of bat fatalities at wind turbines: Current Biology, v. 18, no. 16, p. R695–R696, accessed November 25, 2014, at http://dx.doi. org/10.1016/j.cub.2008.06.029.

Barclay, R.M.R., and Harder, L.D., 2003, Life histories of bats; Life in the slow lane, *in* Kunz, T.H., and Fenton, M.B., eds., Bat ecology: Chicago, Ill., University of Chicago Press, p. 209–256.

Barlow, Jay, Swartz, S.L., Eagle, T.C., and Wade, P.R., 1995, U.S. marine mammal stock assessments; Guidelines for preparation, background, and a summary of the 1995 assessments: U.S. Department of Commerce, National Oceanic and Atmospheric Administration, National Marine Fisheries Service, NOAA Technical Memorandum NMFS–OPR–95– 6, 73 p., accessed November 11, 2014, at http://www.nmfs. noaa.gov/pr/pdfs/sars/guidelines1995.pdf.

Beach, Rocky; Hahn, Deb; Dresler, Paul; Burkett, Chris; Gwynn, Becky; Fuller, Steve; Neel, Larry; Larouche, Genevieve; Regan, Ron; Soule, Judy; Harris, Mike; Gould, Rowan; Figg, Dennis; Haseltine, Susan; Eagle, A.C.; Pairis, Amber; Humpert, Mark; Frame, Mike; Wostlie, Steve; Holt, Lisa; Hess, Karl; Kanter, John; Chadwick, Dave; Hess-Herbert, Janet; and Edelson, Naomi, 2011, State wildlife action plans; Shaping national fish & wildlife conservation: Association of Fish and Wildlife Agencies Report 07–338, 36 p., accessed October 30, 2014, at https://www.dodlegacy. org/Legacy/project/productdocs/National%20Synthesis%20Report-FINAL%20%28DOD%29_bb12c28e-75b2-4fdea6ba-82d288702407.pdf.

Beissinger, S.R., and McCullough, D.R., eds., 2002, Population viability analysis: Chicago, Ill., University of Chicago Press, 593 p.

Bellebaum, J., Korner-Nievergelt, F., Dürr, T., and Mammen, U., 2013, Wind turbine fatalities approach a level of concern in a raptor population: Journal for Nature Conservation, v. 21, no. 6, p. 394–400, accessed November 4, 2013, at http:// dx.doi.org/10.1016/j.jnc.2013.06.001.

Boyce, M.S., 1992, Population viability analysis: Annual Review of Ecology and Systematics, v. 23, p. 481–506. [Also available at http://dx.doi.org/10.1146/annurev. es.23.110192.002405.]

Brennan, S.T., Burruss, R.C., Merrill, M.D., Freeman, P.A., and Ruppert, L.F., 2010, A probabilistic assessment methodology for the evaluation of geologic carbon dioxide storage: U.S. Geological Survey Open-File Report 2010–1127, 31 p., accessed September 22, 2013, at http://pubs.usgs.gov/ of/2010/1127.

Bright, Jenny, Langston, Rowena, Bullman, Rhys, Evans, Richard, Gardner, Sam, and Pearce-Higgins, James, 2008, Map of bird sensitivities to wind farms in Scotland; A tool to aid planning and conservation: Biological Conservation, v. 141, no. 9, p. 2342–2356, accessed June 25, 2013, at http://dx.doi.org/10.1016/j.biocon.2008.06.029.

Burgman, M.A., Ferson, S., and Akçakaya, H.R., 1993, Risk assessment in conservation biology: New York, Chapman and Hall, 314 p.

Burgman, M.A., Keith, D.A., and Walshe, T.V., 1999, Uncertainty in comparative risk analysis for threatened Australian plant species: Risk Analysis, v. 19, no. 4, p. 585–598, accessed December 9, 2013, at http://dx.doi. org/10.1111/j.1539-6924.1999.tb00430.x.

Burnham, K.P., and Anderson, D.R., 1984, Tests of compensatory vs. additive hypotheses of mortality in mallards: Ecology, v. 65, p. 105–112. [Also available at http://dx.doi. org/10.2307/1939463.]

Butler, M.J., Harris, Grant, and Strobel, B.N., 2013, Influence of whooping crane population dynamics on its recovery and management: Biological Conservation, v. 162, p. 89–99, accessed November 4, 2014, at http://dx.doi.org/10.1016/j. biocon.2013.04.003.

Carrete, Martina, Sánchez-Zapata, J.A., Benítez, J.R., Lobón, Manuel, and Donázar, J.A., 2009, Large scale risk-assessment of wind-farms on population viability of a globally endangered long-lived raptor: Biological Conservation, v. 142, p. 2954–2961, accessed September 4, 2014, at http:// dx.doi.org/10.1016/j.biocon.2009.07.027.

Caswell, Hal, 2000, Matrix population models; Construction, analysis, and interpretation: Sunderland, Mass., Sinauer Associates, 722 p.

Charpentier, R.R., and Cook, T.A., 2011, USGS methodology for assessing continuous petroleum resources: U.S. Geological Survey Open-File Report 2011–1167, 73 p., accessed November 3, 2013, at http://pubs.usgs.gov/of/2011/1167/ OF11-1167.pdf.

Coffin, A.W., 2007, From roadkill to road ecology; A review of the ecological effects of roads: Journal of Transport Geography, v. 15, no. 5, p. 396–406, accessed January 21, 2015, at http://dx.doi.org/10.1016/j.jtrangeo.2006.11.006.

Confer, J.L., Serrell, R.E., Hager, Mark, and Lahr, Eleanor, 2008, Field tests of the Rosenberg-Blancher method for converting point counts to abundance estimates: The Auk, v. 125, no. 4, p. 932–938. [Also available at http://dx.doi. org/10.1525/auk.2008.06263.]

Cornell Lab of Ornithology, 2015, eBird: Cornell Lab of Ornithology Web site accessed February 20, 2015, at http:// ebird.org/content/ebird/.

Crockford, N.J., 1992, A review of the possible impacts of windfarms on birds and other wildlife: Joint Nature Conservation Committee Report 27, 65 p.

Crouse, D.T., Crowder, L.B., and Caswell, Hal, 1987, A stage-based population model for loggerhead sea turtles and implications for conservation: Ecology, v. 68, no. 5, p. 1412–1423, accessed November 5, 2014, at http://www. jstor.org/stable/1939225.

Cryan, P.M., 2011, Wind turbines as landscape impediments to the migratory connectivity of bats: Environmental Law, v. 41, p. 355–370, accessed April 26, 2013, at http:// heinonlinebackup.com/hol-cgi-bin/get_pdf.cgi?handle=hein. journals/envlnw41&section=15.

Desholm, Mark, 2009, Avian sensitivity to mortality; Prioritising migratory bird species for assessment at proposed wind farms: Journal of Environmental Management, v. 90, no. 8, p. 2672–2679, accessed December 9, 2014, at http://dx.doi. org/10.1016/j.jenvman.2009.02.005.

Diffendorfer, J.E., Compton, Roger, Kramer, Louisa, Ancona, Zach, and Norton, Donna, 2014, Onshore industrial wind turbine locations for the United States through July 2013: U.S. Geological Survey Data Series 817, accessed February 11, 2014, at http://dx.doi.org/10.3133/ds817.

Dillingham, P.W., and Fletcher, David, 2008, Estimating the ability of birds to sustain additional human-caused mortalities using a simple decision rule and allometric relationships: Biological Conservation, v. 141, no. 7, p. 1783–1792. [Also available at http://dx.doi.org/10.1016/j. biocon.2008.04.022.]

Ebert, T.A., 1998, Plant and animal populations; Methods in demography: San Diego, Calif., Academic Press, 312 p.

Ellison, L.E., 2012, Bats and wind energy; A literature synthesis and annotated bibliography: U.S. Geological Survey Open-File Report 2012–1110, 57 p., accessed April 17, 2013, at http://pubs.usgs.gov/of/2012/1110/OF12-1110.pdf.

Erickson, R.A., Thogmartin, W.E., and Szymanski, J.A., 2014, BatTool: An R package with GUI for assessing the effect of White-nose syndrome and other take events on *Myotis* spp. of bats: Source Code for Biology and Medicine, v. 9, no. 9, 10 p., accessed November 8, 2014, at http://dx.doi.org/10.1186/1751-0473-9-9.

Erickson, W.P., Johnson, G.D., Strickland, M.D., Young, D.P., Jr., Sernka, K.J., and Good, R.E., 2001, Avian collisions with wind turbines; A summary of existing studies and comparisons to other sources of avian collision mortality in the United States: National Wind Coordinating Committee Resource Document, 62 p., accessed October 30, 2014, at http://www.west-inc.com/reports/avian_collisions.pdf.

Erickson, W.P., Wolfe, M.M., Bay, K.J., Johnson, D.H., and Gehring, J.L., 2014, A comprehensive analysis of small-passerine fatalities from collision with turbines at wind energy facilities: PLoS One, v. 9, no. 9, 18 p., accessed October, 3, 2014, at http://dx.doi.org/10.1371/journal. pone. 0107491.

Forman, R.T.T., and Alexander, L.E., 1998, Roads and their major ecological effects: Annual Review of Ecology and Systematics, v. 29, p. 207–231, accessed May 8, 2013, at http://www.jstor.org/stable/221707.

Fox, A.D., Desholm, Mark, Kahlert, Johnny, Christensen, T.K., and Petersen, I.K., 2006, Information needs to support environmental impact assessment of the effects of European marine offshore wind farms on birds: Ibis, v. 148, supplement s1, p. 129–144, accessed June 10, 2013, at http://dx.doi.org/10.1111/j.1474-919X.2006.00510.x.

Frick, W.F., Rainey, W.E., and Pierson, E.D., 2007, Potential effects of environmental contamination on Yuma Myotis demography and population growth: Ecological Applications, v. 17, no. 4, p. 1213–1222, accessed January 6, 2015, at http://frick.eeb.ucsc.edu/wp-content/uploads/2013/11/Ecol-Appl-2007-Frick-1.pdf.

Garthe, Stefan, and Hüppop, Ommo, 2004, Scaling possible adverse effects of marine wind farms on seabirds; Developing and applying a vulnerability index: Journal of Applied Ecology, v. 41, no. 4, p. 724–734, accessed April 30, 2013, at http://dx.doi.org/10.1111/j.0021-8901.2004.00918.x.

Gedamke, Todd, Hoenig, J.M., DuPaul, W.D., and Musick, J.A., 2009, Stock-recruitment dynamics and the maximum population growth rate of the barndoor skate on Georges Bank: North American Journal of Fisheries Management, v. 29, no. 2, p. 512–526, accessed January 21, 2015, at http:// dx.doi.org/10.1577/M07-058.1.

Gedamke, Todd, Hoenig, J.M., Musick, J.A., DuPaul, W.D., and Gruber, S.H., 2007, Using demographic models to determine intrinsic rate of increase and sustainable fishing for elasmobranchs; Pitfalls, advances, and applications: North American Journal of Fisheries Management, v. 27, no. 2, p. 605–618, accessed January 21, 2015, at http:// dx.doi.org/10.1577/ M05-157.1.

Gill, J.P., Townsley, Mike, and Mudge, G.P., 1996, Review of the impacts of wind farms and other aerial structures upon birds: Scottish Natural Heritage Review, no. 21, 68 p., accessed October 14, 2014, at http://www.snh.org.uk/pdfs/ publications/review/021.pdf.

Global Biodiversity Information Facility Secretariat, 2015, Global Biodiversity Information Facility; Free and open access to biodiversity data: Global Biodiversity Information Facility Secretariat Web site accessed January 20, 2015, at http://www.gbif.org.

Grodsky, S.M., Behr, M.J., Gendler, Andrew, Drake, David, Dieterle, B.D., Rudd, R.J., and Walrath, N.L., 2011, Investigating the causes of death for wind turbine-associated bat fatalities: Journal of Mammalogy, v. 92, no. 5, p. 917–925, accessed January 8, 2015, at http://jmammal.oxfordjournals. org/content/jmammal/92/5/917.full.pdf.

Hand, M.M., Baldwin, S., DeMeo, E., Reilly, J.M., Mai, T., Arent, D., Porro, G., Meshek, M., and Sandor, D., eds., 2012, Renewable electricity futures study: National Renewable Energy Laboratory [Report] NREL/TP–6A20–52409, 4 v., accessed November 5, 2014, at http://www.nrel.gov/ analysis/re_futures.

Hayes, M.A., 2013, Bats killed in large numbers at United States wind energy facilities: BioScience, v. 63, no. 12, p. 975–979, accessed January 21, 2015, at http://dx.doi. org/10.1525/bio.2013.63.12.10.

Huso, M.M.P., and Dalthorp, Dan, 2014, A comment on "Bats killed in large numbers at United States wind energy facilities": BioScience, v. 64, no. 6, p. 546–547, accessed January 21, 2015, at http://dx.doi.org/10.1093 /biosci/biu056.

International Energy Agency, 2012, World energy outlook 2012; International Energy Agency, 690 p., accessed October 2014, at http://dx.doi.org/10.1787/weo-2012-en. [Also available at http://www. iea.org/publications/freepublications/ publication/world-energy-outlook-2012.html.]

International Union for Conservation of Nature, 2012, IUCN Red List categories and criteria, version 3.1 (2d ed.): Gland, Switzerland, and Cambridge, United Kingdom, International Union for Conservation of Nature, 32 p. [Also available at http://jr.iucnredlist.org/documents/ redlist_cats_ crit_en.pdf.]

International Union for Conservation of Nature, 2014, The IUCN Red List of Threatened Species, version 2014.3: International Union for Conservation of Nature Web site, accessed October 31, 2014, at www.iucnredlist.org.

Johnston, N.N., Bradley, J.E., and Otter, K.A., 2014, Increased flight altitudes among migrating golden eagles suggest turbine avoidance at a Rocky Mountain wind installation: PLoS ONE, v. 9, no. 3, accessed November 4, 2014, at http://dx.doi.org/10.1371/journal.pone.0093030.

Katzner, T., Johnson, J.A., Evans, D.M., Garner, T.W.J., Gompper, M.E., Altwegg, R., Branch, T.A., Gordan, I.J., and Pettorelli, N., 2013, Challenges and opportunities for animal conservation from renewable energy development: Animal Conservation, v. 16, no. 4, p. 367–369, accessed January 8, 2015, at http://dx.doi.org/10.1111/acv.12067.

Keith, D.A., McCarthy, M.A., Regan, Helen, Regan, Tracey, Bowles, Christy, Drill, Claire, Craig, Corey, Pellow, Belinda, Burgman, M.A., Master, L.L., Ruckelshaus, Mary, Mackenzie, Berin, Andelman, S.J., and Wade, P.R., 2004, Protocols for listing threatened species can forecast extinction: Ecology Letters, v. 7, no. 11, p. 1101–1108, accessed December 9, 2013, at http://dx.doi.org/10.1111/j.1461- 0248.2004.00663.x.

Kingsley, Andrea, and Whittam, Becky, 2005, Wind turbines and birds; A background review for environmental assessment: Gatineau, Quebec, Canada, Environment Canada, Canadian Wildlife Service, 81 p. [Also available at http:// www.energy.ca.gov/windguidelines/documents/other_guidelines/2006-05-12_BCKGRD_ENVIRMTL_ASSMNT. PDF.]

Kitano, Masato, and Shiraki, Saiko, 2013, Estimation of bird fatalities at wind farms with complex topography and vegetation in Hokkaido, Japan: Wildlife Society Bulletin, v. 37, no. 1, p. 41–48, accessed December 9, 2013, at http:// dx.doi.org/10.1002/wsb.255.

Koons, D.N., Grand, J.B., Zinner, Bertram, and Rockwell, R.F., 2005, Transient population dynamics; Relations to life history and initial population state: Ecological Modelling, v. 185, p. 283–297. [Also available at http://www.demogr. mpg.de/publications/files/2364_1155281974_1_Koons%20 et%20al%202005.pdf.]

Kushlan, J.A.; Steinkamp, M.J.; Parsons, K.C.; Capp, Jack; Acosta Cruz, Martin; Coulter, Malcolm; Davidson, Ian; Dickson, Loney; Edelson, Naomi; Elliot, Richard; Erwin, R.M.; Hatch, Scott; Kress, Stephen; Milko, Robert; Miller, Steve; Mills, Kyra; Paul, Richard; Phillips, Roberto; Saliva, J.E.; Sydeman, Bill; Trapp, John; Wheeler, Jennifer; and Wohl, Kent, 2002, Waterbird conservation for the Americas; The North American waterbird conservation plan, version 1: Washington, D.C., Waterbird Conservation for the Americas, 78 p. [Version 1 and updates are available at http://www.waterbirdconservation.org/nawcp.html.]

Kuvlesky, W.P., Jr., Brennan, L.A., Morrison, M.L., Boydston, K.K., Ballard, B.M., and Bryant, F.C., 2007, Wind energy development and wildlife conservation; Challenges and opportunities: Journal of Wildlife Management, v. 71, no. 8, p. 2487–2498, accessed June 10, 2013, at http://dx.doi. org/10.2193/2007-248.

Langston, R.H.W., and Pullan, J.D., 2004, Effects of wind farms on birds: Council of Europe Publishing, Nature and Environment no. 139, 89 p., accessed November 4, 2014, at http://www.coe.int/t/dg4/cultureheritage/nature/bern/documents/072011_IKB/Effects_wind_farms_on_birds.pdf.

Leddy, K.L., Higgins, K.F., and Naugle, D.E., 1999, Effects of wind turbines on upland nesting birds in Conservation Reserve Program grasslands: Wilson Bulletin, v. 111, no. 1, p. 100–104.

Levins, Richard, 1966, The strategy of model building in population biology: American Scientist, v. 54, no. 4, p. 421–431.

Loeb, S.C., Rodhouse, T.J., Ellison, L.E., Lausen, C.L., Reichard, J.D., Irvine, K.M., Ingersoll, T.E., Coleman, J.T.H., Thogmartin, W.E., Sauer, J.R., Francis, C.M., Bayless, M.L., Stanley, T.R., and Johnson, D.H., in press, A plan for the North American Bat Monitoring Program (NABat): U.S. Department of Agriculture, Forest Service, Southern Research Station, General Technical Report.

Lonergan, Mike, 2011, Potential biological removal and other currently used management rules for marine mammal populations; A comparison: Marine Policy, v. 35, no. 5, p. 584–589, accessed November 25, 2014, at http://www. sciencedirect.com/science/article/pii/S0308597X11000212.

Loss, S.R., Will, Tom, and Marra, P.P., 2013, Estimates of bird collision mortality at wind facilities in the contiguous United States: Biological Conservation, v. 168, p. 201–209, accessed April 17, 2013, at http://dx.doi.org/10.1016/j. biocon.2013.10.007.

Martin, T.G., Burgman, M.A., Fidler, Fiona, Kuhnert, P.M., Low-Choy, Samantha, McBride, Marissa, and Mengersen, Kerrie, 2012, Eliciting expert knowledge in conservation science: Conservation Biology, v. 26, no. 1, p. 29–38, accessed January 21, 2015, at http://dx.doi.org/10.1111/j.1523-1739.2011.01806.x.

McCracken, G.F., 2003, Estimates of population sizes in summer colonies of Brazilian free-tailed bats (*Tadarida brasiliensis*), *in* O'Shea, T.J., and Bogan, M.A., eds., Monitoring trends in bat populations of the United States and territories; Problems and prospects: U.S. Geological Survey, Biological

Resources Discipline, Information and Technology Report USGS/BRD/ITR–2003–003, p. 21–30. [Also available at https://www.fort.usgs.gov/sites/default/files/products/publications/21329/21329.pdf.]

Michel, N., DeSante, D.F., Kaschube, D.R., and Nott, M.P., 2006, The Monitoring Avian Productivity and Survivorship (MAPS) Program annual reports, 1989–2003: Institute for Bird Populations / NBII bird conservation node, NBII/ MAPS Avian Demographics Query Interface, accessed November 5, 2014, at http://www.birdpop.org/nbii2006/nbiihome.asp.

Millar, R.B., and Meyer, Renate, 2000, Non-linear state space modelling of fisheries biomass dynamics by using Metropolis-Hastings within-Gibbs sampling: Journal of the Royal Statistical Society: Series C (Applied Statistics), v. 49, no. 3, p. 327–342, accessed January 21, 2015, at http://dx.doi.org/10.1111/1467-9876.00195.

Miller, T.A., Brooks, R.P., Lanzone, Michael, Brandes, David, Cooper, Jeff, O'Malley, Kieran, Maisonneuve, Charles, Tremblay, Junior, Duerr, Adam, and Katzner, Todd, 2014, Assessing risk to birds from industrial wind energy development via paired resource selection models: Conservation Biology, v. 28, no. 3, p. 745–755, accessed January 8, 2015, at http://dx.doi.org/10.1111/cobi.12227.

Milner-Gulland, E.J., and Akçakaya, H.R., 2001, Sustainability indices for exploited populations: Trends in Ecology & Evolution, v. 16, no. 12, p. 686–692, accessed January 21, 2015, at http://dx.doi.org/10.1016/S0169-5347(01)02278-9.

Moore, J.E., Curtis, K.A., Lewison, R.L., Dillingham, P.W., Cope, J.M., Fordham, S.V., Heppell, S.S., Pardo, S.A., Simpfendorfer, C.A., Tuck, G.N., and Zhou, S., 2013, Evaluating sustainability of fisheries bycatch mortality for marine megafauna; A review of conservation reference points for data-limited populations: Environmental Conservation, v. 40, no. 4, p. 329–344, accessed January 21, 2015, at http://dx.doi.org/ 10.1017/S037689291300012X.

Morris, W.F., and Doak, D.F., 2002, Quantitative conservation biology; Theory and practice of population viability analysis: Sunderland, Mass., Sinauer Associates, 480 p.

Munday, P.L., 2004, Habitat loss, resource specialization, and extinction on coral reefs: Global Change Biology, v. 10, no. 10, p. 1642–1647, accessed October 30, 2014, at http:// dx.doi.org/10.1111/j.1365-2486.2004.00839.x.

National Audubon Society, 2015, Christmas Bird Count: National Audubon Society Web page, accessed January 6, 2015, at http://www.audubon.org/ conservation/science/ christmas-bird-count.

National Marine Fisheries Service, 2015, Guidelines for assessing marine mammal stocks (GAMMS): National Marine Fisheries Service Web page, accessed February 5, 2015, at http://www.nmfs.noaa.gov/pr/ sars/guidelines.htm.

National Research Council, Committee on Environmental Impacts of Wind-Energy Projects, 2007, Environmental impacts of wind-energy projects: Washington, D.C., National Academies Press, 376 p., accessed April 26, 2013, at http://www.nap.edu/catalog/11935/environmentalimpacts-of-wind-energy-projects.

Nichols, J.D., Conroy, M.J., Anderson, D.R., and Burnham, K.P., 1984, Compensatory mortality in waterfowl populations; A review of the evidence and implications for research and management: Transactions of the North American Wildlife and Natural Resources Conference, v. 49, p. 535–554.

Niel, Colin, and Lebreton, J.-D., 2005, Using demographic invariants to detect overharvested bird populations from incomplete data: Conservation Biology, v. 19, no. 3, p. 826–835, accessed January 21, 2015, at http://dx.doi.org/ 10.1111/j.1523-1739.2005.00310.x.

O'Grady, J.J., Burgman, M.A., Keith, D.A., Master, L.L., Andelman, S.J., Brook, B.W., Hammerson, G.A., Regan, Tracey, and Frankham, Richard, 2004, Correlations among extinction risks assessed by different systems of threatened species categorization: Conservation Biology, v. 18, no. 6, p. 1624–1635, accessed December 9, 2013, at http://dx.doi. org/10. 1111/j.1523-1739.2004.00109.x.

O'Shea, T.J., Ellison, L.E., and Stanley, T.R., 2003, Survival estimation in bats; Historical overview, critical appraisal, and suggestions for new approaches, *in* Thompson, William, ed., Sampling rare or elusive species; Concepts, designs, and techniques for estimating population parameters: Washington, D.C., Island Press, p. 297–336.

Oli, M.K., 2004, The fast-slow continuum and mammalian life-history patterns; An empirical evaluation: Basic and Applied Ecology, v. 5, no. 5, p. 449–463, accessed October 30, 2014, at http://dx.doi.org/10. 1016/j.baae.2004.06.002.

Pagel, J.E., Kritz, K.J., Millsap, B.A., Murphy, R.K., Kershner, E.L., and Covington, Scott, 2013, Bald eagle and golden eagle mortalities at wind energy facilities in the contiguous United States: Journal of Raptor

Research, v. 47, no. 3, p. 311–315, accessed September 12, 2013, at http://dx.doi.org/10.3356/JRR-12-00019.1.

Partners in Flight Science Committee, 2013, Population estimates database, version 2013: accessed November 5, 2014, at http://rmbo.org/pifpopestimates.

Pearse, A.T., and Selbo, S.M., 2012, Model of whooping crane energetics as foundation for development of a method to assess potential take during migration: U.S. Geological Survey Open-File Report 2012–1156, 13 p., accessed November 4, 2014, at http://pubs.usgs.gov/of/2012/1156/ of12-1156.pdf.

Percival, Steve, 2005, Birds and windfarms; What are the real issues?: British Birds, v. 98, p. 194–204, accessed November 4, 2014, at http://britishbirds.co.uk/wp-content/uploads/ article_files/V98/V98_N04/V98_N04_P194_204_A002.pdf.

Péron, Guillaume, Hines, J.E., Nichols, J.D., Kendall, W.L., Peters, K.A., and Mizrahi, D.S., 2013, Estimation of bird and bat mortality at wind-power farms with superpopulation models: Journal of Applied Ecology, v. 50, no. 4, p. 902–911, accessed February 12, 2014, at http://dx.doi.org/10.1111/1365-2664.12100.

Phillips, Sue, 2011, Wind energy and wildlife research at the Forest and Rangeland Ecosystem Science Center: U.S. Geological Survey Fact Sheet 2011–3134, 4 p., accessed on November 25, 2014, at http://pubs.usgs.gov/fs/2011/3134/.

Poole, A., ed., 2005, The birds of North America online: Cornell Lab of Ornithology, accessed October 20, 2014, at http://bna.birds.cornell.edu/bna/.

Rich, T.D., Beardmore, C.J., Berlanga, H., Blancher, P.J., Bradstreet, M.S.W., Butcher, G.S., Demarest, D.W., Dunn, E.H., Hunter, W.C., Iñigo-Elias, E.E., Kennedy, J.A., Martell, A.M., Panjabi, A.O., Pashley, D.N., Rosenberg, K.V., Rustay, C.M., Wendt, J.S., and Will, T.C., 2004, Partners in Flight North American Landbird Conservation Plan: Cornell Lab of Ornithology, accessed November 21, 2014, at http://www.partnersinflight.org/cont_plan.

Richard, Y., and Abraham, E.R., 2013, Risk of commercial fisheries to New Zealand seabird populations: New Zealand Aquatic Environment and Biodiversity Report No. 109, 62 p., accessed January 25, 2015, at http://fs.fish.govt.nz/Doc/23121/AEBR_109_2596_PRO2010-02,%20Obj.%20 1,%20MS4,%20RR2,1.pdf.ashx.

Rollins, K.E., Meyerholz, D.K., Johnson, G.D., Capparella, A.P., and Loew, S.S., 2012, A forensic investigation into the etiology of bat mortality at a wind farm; Barotrauma or traumatic injury?: Veterinary Pathology, v. 49, no. 2, p. 362–371, accessed January 8, 2015, at http://dx.doi.org/10.1177/0300985812436745.

Rose, K.A., and Cowan, J.H., Jr., 2003, Data, models, and decisions in U.S. marine fisheries management; Lessons for ecologists: Annual Review of Ecology, Evolution, and Systematics, v. 34, p. 127–151.

Runge, M.C., Converse, S.J., and Lyons, J.E., 2011, Which uncertainty? Using expert elicitation and expected value of information to design an adaptive program: Biological Conservation, v. 144, no. 4, p. 1214–1223, accessed January 21, 2015, at http://dx.doi.org/10.1016/j.biocon.2010.12.020.

Runge, M.C., Sauer, J.R., Avery, M.L., Blackwell, B.F., and Koneff, M.D., 2009, Assessing allowable take of migratory birds: Journal of Wildlife Management, v. 73, no. 4, p. 556–565.

Russell, R.E., Tinsley, Karl, Erickson, R.A., Thogmartin, W.E., and Szymanski, Jennifer, 2014, Estimating the spatial distribution of wintering little brown bat populations in the eastern United States: Ecology and Evolution, v. 4, no. 19, p. 3746–3754, accessed November 18, 2014, at http://dx.doi.org/10.1002/ece3.1215.

Rydell, Jens, Bach, Lothar, Dubourg-Savage, M.-J., Green, Martin, Rodrigues, Luisa, and Hedenström, Anders, 2010, Bat mortality at wind turbines in northwestern Europe: Acta Chiropterologica, v. 12, no. 2, p. 261–274, accessed December 9, 2014, at http://dx.doi.org/10.3161/15 081101 0X537846.

Sauer, J.R., Hines, J.E., Fallon, J.E., Pardieck, K.L., Ziolkowski, D.J., Jr., and Link, W.A., 2015, The North American Breeding Bird Survey, results and analysis 1966– 2013, version 01.30.2015: U.S. Geological Survey Web site, accessed February 5, 2015, at http://www.mbr-pwrc.usgs. gov/bbs/.

Schaub, Michael, 2012, Spatial distribution of wind turbines is crucial for the survival of red kite populations: Biological Conservation, v. 155, p. 111–118, accessed June 12, 2013, at http://dx.doi.org/10.1016/ j.biocon. 2012.06.021.

Schmoker, J.W., and Klett, T.R., 2005, U.S. Geological Survey assessment concepts for conventional petroleum accumulations, chap. 19 *of* USGS Southwestern Wyoming Province Assessment Team, Petroleum systems and geologic assessment of oil and gas in the Southwestern Wyoming Province, Wyoming, Colorado, and Utah: U.S. Geological Survey Digital Data Series DDS–69–D, 9 p., accessed December 10, 2014, at http://pubs.usgs.gov/dds/dds-069/dds-069-d/ REPORTS/69_D_CH_19.pdf.

Sharp, D.E., Kruse, K.L., and Thorpe, P.P., 2002, The midwinter waterfowl survey in the central flyway: U.S. Fish and Wildlife Service, Division of Migratory Bird Management, accessed November 5, 2014, at https://migbirdapps.fws. gov/ as https://migbirdapps.fws.gov/ mbdc databases/mwi/ CF_MWS_Assessment_2002.pdf.

Slade, N.A., Gomulkiewicz, Richard, and Alexander, H.M., 1998, Alternatives to Robinson and Redford's method of assessing overharvest from incomplete demographic data: Conservation Biology, v. 12, no. 1, p. 148–155, accessed January 21, 2015, at http://dx.doi.org/10.1111/j.1523-1739.1998.96273.x.

Smallwood, K.S., 2013, Comparing bird and bat fatality-rate estimates among North American wind-energy projects: Wildlife Society Bulletin, v. 37, no. 1, p. 19–33, accessed May 9, 2013, at http://dx.doi.org/10.1002/wsb.260.

Smallwood, K.S., and Karas, Brian, 2009, Avian and bat fatality rates at old-generation and repowered wind turbines in California: Journal of Wildlife Management, v. 73, no. 7, p. 1062–1071, accessed April 17, 2013, at http://dx.doi. org/10.2193/2008-464.

Stahl, J.T., and Oli, M.K., 2006, Relative importance of avian life-history variables to population growth rate: Ecological Modelling, v. 198, nos. 1–2, p. 23–39, accessed April 17, 2013, at http://dx.doi.org/10. 1016/j. ecolmodel.2006.04.001.

Stevens, T.K., Hale, A.M., Karsten, K.B., and Bennett, V.J., 2013, An analysis of displacement from wind turbines in a wintering grassland bird community: Biodiversity and Conservation, v. 22, no. 8, p. 1755–1767, accessed November 24, 2014, at http://dx.doi.org/10.1007/s10531-013-0510-8.

Strickland, M.D., Arnett, E.B., Erickson, W.P., Johnson, D.H., Johnson, G.D., Morrison, M.L., Shaffer, J.A., and Warren-Hicks, William, 2011, Comprehensive guide to studying wind energy/wildlife interactions: Washington, D.C., National Wind Coordinating Collaborative, 281 p., accessed April 17, 2013, at http://www.batcon.org/pdfs/wind/ National%20Wind%20Coordinating%20Collaborative%20 2011_ Compre hensive%20Guide%20to%20Studying%20 Wind%20Energy%20and%20Wildlife%20Interactions.pdf.

Swihart, R.K., Gehring, T.M., Kolozsvary, M.B., and Nupp, T.E., 2003, Responses of 'resistant' vertebrates to habitat loss and fragmentation; The importance of niche breadth and range boundaries: Diversity and Distributions, v. 9, no. 1, p. 1–18, accessed December 9, 2013, at http://dx.doi. org/10.1046/j.1472-4642.2003.00158.x.

Taylor, B.L., 1993, "Best" abundance estimates and best management; Why they are not the same: U.S. Department of Commerce, National Oceanic and Atmospheric Administration, National Marine Fisheries Service, NOAA Technical Memorandum NMFS–SWFSC–188, 20 p. [Also available at https://swfsc.noaa.gov/publications/TM/SWFSC/NOAATM-NMFS-SWFSC-188.PDF.]

Taylor, B.L., Wade, P.R., De Master, D.P., and Barlow, Jay, 2000, Incorporating uncertainty into management models for marine mammals: Conservation Biology, v. 14, no. 5, p. 1243–1252, accessed January 21, 2015, at http://dx.doi. org/10.1046/j.1523-1739.2000.99409.x.

Thogmartin, W.E., King, R.A., McKann, P.C., Szymanski, J.A., and Pruitt, Lori, 2012, Population-level impact of white-nose syndrome on the endangered Indiana bat: Journal of Mammalogy, v. 93, no. 4, p. 1086–1098, accessed November 4, 2014, at http://dx.doi.org/10.1644/11-MAMM-A-355.1.

Thogmartin, W.E., Sanders-Reed, C.A., Szymanski, J.A., McKann, P.C., Pruitt, Lori, King, R.A., Runge, M.C., and Russell, R.E., 2013, White-nose syndrome is likely to extirpate the endangered Indiana bat over large parts of its range: Biological Conservation, v. 160, p. 162–172, accessed November 18, 2014, at http://www.sciencedirect.com/science/article/pii/S0006320713000207.

U.S. Department of the Interior, 2014, Powering our future, *in* Departmental highlights, *in* Fiscal year 2015, the Interior budget in brief: U.S. Department of the Interior, p. DH–21 to DH–24, accessed November 4, 2014, at http://www.doi. gov/budget/appropriations/ 2015/highlights/upload/DH021. pdf.

U.S. Department of the Interior and U.S. Department of Agriculture, 2011, New energy frontier; Balancing energy development on Federal lands—A joint report to Congress on siting energy development projects on Federal lands: U.S. Department of the Interior and U.S. Department of Agriculture report, 138 p., accessed February 5, 2015, at http://www.doi.gov//whatwedo/energy/index.cfm.

U.S. Energy Information Administration, 2014a, Annual energy outlook 2014, with projections to 2040: U.S. Energy Information Administration [Report] DOE/ EIA–0383(2014), variously paged [269 p.], accessed November 10, 2014, at http://www.eia.gov/forecasts/aeo/pdf/0383(2014).pdf.

U.S. Energy Information Administration, 2014b, Monthly energy review, October 2014: U.S. Energy Information Administration Monthly Energy

Review DOE/ EIA–0035(2014/10), 201 p., accessed November 5, 2014, at http://www.eia.gov/totalenergy/data/monthly/ archive/00351410.pdf.

U.S. Fish and Wildlife Service, 2008, Birds of conservation concern 2008: U.S. Fish and Wildlife Service, Division of Migratory Bird Management, 85 p., accessed November 11, 2014, at http://www.fws.gov/migratorybirds/NewReportsPublications/SpecialTopics/BCC2008/BCC 2008.pdf.

U.S. Fish and Wildlife Service, 2011a, Draft eagle conservation plan guidance: U.S. Fish and Wildlife Service, 106 p., accessed October 5, 2014, at http://www.fws.gov/windenergy/docs/ECP_draft_guidance_2j0_final_clean_omb.pdf.

U.S. Fish and Wildlife Service, 2011b, U.S. Fish and Wildlife Service draft land-based wind energy guidelines: U.S. Fish and Wildlife Service, 61 p., plus appendixes A–C, accessed October 5, 2014, at http://www.fws.gov/windenergy/docs/ Final_Wind_Energy_Guidelines_2_8_11_ CLEAN.pdf.

U.S. Fish and Wildlife Service, 2012, U.S. Fish and Wildlife Service land-based wind energy guidelines: U.S. Fish and Wildlife Service, 71 p., accessed October 5, 2014, at http:// www.fws.gov/windenergy/docs/WEG_final.pdf.

U.S. Fish and Wildlife Service, 2014, Endangered species: U.S. Fish and Wildlife Service Web site, accessed November 4, 2014, at http://www.fws.gov/endangered/species/ us-species.html.

U.S. Geological Survey, 2015, North American Bat Monitoring Program (NABat): U.S. Geological Survey Web page, accessed February 5, 2015, at https://www.fort.usgs.gov/ science-tasks/2457.

Voigt, C.C., Popa-Lisseanu, A.G., Niermann, Ivo, and KramerSchadt, Stephanie, 2012, The catchment area of wind farms for European bats; A plea for international regulations: Biological Conservation, v. 153, p. 80–86, accessed January 8, 2015, at http://www.sciencedirect.com/science/article/pii/ S0006320712002091.

Wade, P.R., 1998, Calculating limits to the allowable human-caused mortality of cetaceans and pinnipeds: Marine Mammal Science, v. 14, no. 1, p. 1–37. [Also available at http:// dx.doi.org/10.1111/j.1748-7692.1998.tb00688.x.]

Watson, J.W., Duff, A.A., and Davies, R.W., 2014, Home range and resource selection by GPS-monitored adult golden eagles in the Columbia Plateau Ecoregion; Implications for wind power development: Journal of Wildlife Management, v. 78, no. 6, p. 1012–1021, accessed November 4, 2014, at http://dx.doi.org/10.1002/jwmg.745.

Wilson, D.E., and Reeder, D.M., eds., 2005, Mammal species of the world; A taxonomic and geographic reference (3d ed.): Baltimore, Md., Johns Hopkins University Press, 2 v., 2142 p.

Winder, V.L., Gregory, A.J., McNew, L.B., and Sandercock, B.K., in press, Responses of male Greater Prairie-Chickens to wind energy development; The Condor, v. 117, no. 2, http://dx.doi.org/10.1650/CONDOR-14-98.1.

Zimmerling, J.R., Pomeroy, A.C., d'Entremont, M.V., and Francis, C.M., 2013, Canadian estimate of bird mortality due to collisions and direct habitat loss associated with wind turbine developments: Avian Conservation and Ecology, v. 8, no. 2, article 10, accessed January 6, 2014, at http:// dx.doi.org/10.5751/ACE-00609-080210.

REFERENCES CITED FOR APPENDIX 1

Arnett, E.B., and Baerwald, E.F., 2013, Impacts of wind energy development on bats; Implications for conservation, in Adams, R.A., and Pedersen, S.C., eds., Bat evolution, ecology, and conservation: New York, Springer, p. 435–456, accessed on January 15, 2015, at http://link.springer.com/chapter/10.1007/978-1-4614-7397-8_21.

Arnett, E.B., Brown, W.K., Erickson, W.P., Fiedler, J.K., Hamilton, B.L., Henry, T.H., Jain, Aaftab, Johnson, G.D., Kerns, Jessica, Koford, R.R., Nicholson, C.P., O'Connell, T.J., Piorkowski, M.D., and Tankersley, R.D., Jr., 2008, Patterns of bat fatalities at wind energy facilities in North America: Journal of Wildlife Management, v. 72, no. 1, p. 61–78, accessed May 14, 2013, at http://dx.doi.org/10.2193/2007-221.

Blumstein, D.T., Mennill, D.J., Clemins, Patrick, Girod, Lewis, Yao, Kung, Patricelli, Gail, Deppe, J.L., Krakauer, A.H., Clark, Christopher, Cortopassi, K.A., Hanser, S.F., McCowan, Brenda, Ali, A.M., and Kirschel, A.N.G., 2011, Acoustic monitoring in terrestrial environments using microphone arrays; Applications, technological considerations and prospectus: Journal of Applied Ecology, v. 48, no. 3, p. 758–767, accessed November 4, 2014, at http:// dx.doi.org/10.1111/j.1365-2664.2011.01993.x.

Britzke, E.R., Gillam, E.H., and Murray, K.L., 2013, Current state of understanding of ultrasonic detectors for the study of bat ecology: Acta Theriologica, v. 58, no. 2, p. 109–117, accessed November 5, 2014, at http://dx.doi.org/10.1007/ s13364-013-0131-3.

Cornell Lab of Ornithology, 2015, eBird: Cornell Lab of Ornithology Web site, accessed February 20, 2015, at http:// ebird.org/content/ebird/.

Diehl, R.H., and Larkin, R.P., 2005, Introduction to the WSR–88D (NEXRAD) for ornithological research, in Ralph, C.J., and Rich, T.D., eds., Bird conservation implementation and integration in the Americas; Proceedings of the Third International Partners in Flight Conference [Asilomar Conference Center in Pacific Grove, California], March 20–24, 2002: U.S. Department of Agriculture, Forest Service, Pacific Southwest Research Station, General Technical Report PSW–GTR–191, p. 876–888. [Also available at http://www.fs.fed.us/psw/publications/documents/psw_gtr191/Asilomar/toc.html.]

Diffendorfer, J.E., Compton, Roger, Kramer, Louisa, Ancona, Zach, and Norton, Donna, 2014, Onshore industrial wind turbine locations for the United States through July 2013: U.S. Geological Survey Data Series 817, accessed February 11, 2014, at http://dx.doi.org/10.3133/ds817.

Duerr, A.E., Miller, T.A., Lanzone, Michael, Brandes, Dave, Cooper, Jeff, O'Malley, Kieran, Maisonneuve, Charles, Tremblay, Junior, and Katzner, Todd, 2012, Testing an emerging paradigm in migration ecology shows surprising differences in efficiency between flight modes: PLoS ONE, v. 7, no. 4, article e35548, 7 p., accessed January 8, 2015, at http://dx.doi.org/10.1371/journal.pone.0035548.

Erickson, R.A., Thogmartin, W.E., Russell, R.E., Diffendorfer, J.E., and Szymanski, J.A., 2014, A stage-structured, spatially explicit migration model for Myotis bats; Mortality location affects system dynamics: Letters in Biomathematics, v. 1, no. 2, p. 157–172, accessed November 8, 2014, at http://www. tandfonline.com/doi/pdf/10.1080/23737867.2014. 11414477.

Erickson, R.A., Thogmartin, W.E., and Szymanski, J.A., 2014, BatTool: An R package with GUI for assessing the effect of White-nose syndrome and other take events on Myotis spp. of bats: Source Code for Biology and Medicine, v. 9, no. 9, 10 p., accessed November 8, 2014, at http://dx.doi. org/10.1186/1751-0473-9-9.

Hayes, M.A., 2013, Bats killed in large numbers at United States wind energy facilities: BioScience, v. 63, no. 12, p. 975–979, accessed January 21, 2015, at http://dx.doi.org/ 10.1525/bio.2013.63.12.10.

Huso, M.M.P., and Dalthorp, Dan, 2014, A comment on "Bats killed in large numbers at United States wind energy facilities": BioScience, v. 64, no. 6, p. 546–547, accessed January 21, 2015, at http://dx.doi.org/ 10.1093/ biosci/biu056.

Katzner, T.E., Brandes, David, Miller, Tricia, Lanzone, Michael, Maisonneuve, Charles, Tremblay, J.A., Mulvihill, Robert, and Merovich, G.T., Jr., 2012, Topography drives migratory flight altitude of golden eagles; Implications for on-shore wind energy development: Journal of Applied Ecology, v. 49, no. 5, p. 1178–1186, accessed January 8, 2015, at http://dx.doi.org/10.1111/j.1365- 2664.2012.02185.x.

Kunz, T.H., Gauthreaux, S.A., Hristov, N.I., Horn, J.W., Jones, Gareth, Kalko, E.K.V., Larkin, R.P., McCracken, G.F., Swartz, S.M., Srygley, R.B., Dudley, Robert, Westbrook, J.K., and Wikelski, Martin, 2008, Aeroecology; Probing and modeling the aerosphere: Integrative and Comparative Biology, v. 48, no. 1, p. 1–11, accessed November 4, 2014, at http://dx.doi.org/10.1093/icb/icn037.

Lanzone, M.J., Miller, T.A., Turk, P., Brandes, D., Halverson, C., Maisonneuve, C., Tremblay, J., Cooper, J., O'Malley, K., Brooks, R.P., and Katzner, T., 2012, Flight responses by a migratory soaring raptor to changing meteorological conditions: Biology Letters, v. 8, no. 5, p. 710–713, accessed January 8, 2015, at http://dx.doi.org/ 10.1098/ rsbl.2012.0359.

Loss, S.R., Will, Tom, and Marra, P.P., 2013, Estimates of bird collision mortality at wind facilities in the contiguous United States: Biological Conservation, v. 168, p. 201–209, accessed April 17, 2013, at http://dx.doi.org/10.1016/j. biocon.2013.10.007.

Miller, T.A., Brooks, R.P., Lanzone, Michael, Brandes, David, Cooper, Jeff, O'Malley, Kieran, Maisonneuve, Charles, Tremblay, Junior, Duerr, Adam, and Katzner, Todd, 2014, Assessing risk to birds from industrial wind energy development via paired resource selection models: Conservation Biology, v. 28, no. 3, p. 745–755, accessed January 8, 2015, at http://dx.doi.org/10.1111/cobi.12227.

Old Bird, Inc., 2015, Nocturnal Flight Call Activity Index: Old Bird Web page, accessed February 25, 2015, at http://www. oldbird.org/.

Sauer, J.R., Hines, J.E., Fallon, J.E., Pardieck, K.L., Ziolkowski, D.J., Jr., and Link, W.A., 2015, The North American Breeding Bird Survey, results and analysis 1966– 2013, version 01.30.2015: U.S. Geological Survey Web site, accessed February 5, 2015, at http://www.mbr-pwrc.usgs. gov/bbs/.

Smallwood, K.S., 2013, Comparing bird and bat fatality-rate estimates among North American wind-energy projects: Wildlife Society Bulletin, v. 37, no. 1, p. 19–33, accessed May 9, 2013, at http://dx.doi.org/ 10.1002/wsb.260.

Stutchbury, B.J.M., Tarof, S.A., Done, Tyler, Gow, Elizabeth, Kramer, P.M., Tautin, John, Fox, J.W., and Afanasyev, Vsevolod, 2009, Tracking long-distance songbird migration by using geolocators: Science, v. 323, no. 5916, p. 896, accessed November 4, 2014, at http://dx.doi.org/10.1126/science.1166664.

Thogmartin, W.E., King, R.A., McKann, P.C., Szymanski, J.A., and Pruitt, Lori, 2012, Population-level impact of white-nose syndrome on the endangered Indiana bat: Journal of Mammalogy, v. 93, no. 4, p. 1086–1098, accessed November 4, 2014, at http://dx.doi.org/10.1644/11-MAMM-A-355.1.

Zimmo, S., Blanco, J., and Nebel, S., 2012, The use of stable isotopes in the study of animal migration. Nature Education Knowledge, v. 3, no. 12, 3 p., accessed on November 4, 2014, at http://www.nature.com/scitable/knowledge/library/ the-use-of-stable-isotopes-in-the-96648168.

REFERENCES CITED FORPENDIX 2

Conroy, M.J., Cohen, Yosef, James, F.C., Matsinos, Y.G., and Maurer, B.A., 1995, Parameter estimation, reliability, and model improvement for spatially explicit models of animal populations: Ecological Applications, v. 5, p. 17–19. [Also available at http://dx.doi.org/10.2307/1942047.]

Holmes, E.E., Sabo, J.L., Viscido, S.V., and Fagan, W.F., 2007, A statistical approach to quasi-extinction forecasting: Ecology Letters, v. 10, no. 12, p. 1182–1198, accessed October 29, 2014, at http://dx.doi. org/10.1111/j.1461- 0248.2007.01105.x.

Sauer, J.R., Hines, J.E., Fallon, J.E., Pardieck, K.L., Ziolkowski, D.J., Jr., and Link, W.A., 2015, The North American Breeding Bird Survey, results and analysis 1966– 2013, version 01.30.2015: U.S. Geological Survey Web site, accessed February 5, 2015, at http://www.mbr-pwrc.usgs. gov/bbs/.

End Notes

[1] U.S. Geological Survey.

[2] Oklahoma State University, Department of Natural Resource Ecology and Management.

[3] University of Minnesota, Department of Fisheries, Wildlife, and Conservation Biology.

INDEX

#

21st century, 90

A

abundance data, 103, 151
access, 156
accounting, 50
activity level, 9, 10, 11, 22, 23, 33, 38, 43, 62, 67, 68
activity rate, 22, 23, 24
adjustment, 130
adults, 35, 58
adverse effects, 55, 138, 155
age, 58, 92, 111, 116, 117, 131, 143, 144, 149
agencies, 8, 59, 95, 97, 102, 106, 107, 124, 130, 131, 143
agriculture, 70
Air Force, 72
Alaska, 84, 101
amputation, 52
appropriations, 164
Assateague Island National Seashore, 13, 49
atelectasis, 36
avian, 10, 58, 65, 66, 68, 75, 78, 81, 82, 84, 85, 87, 88, 89, 109, 127, 129, 131, 148, 155, 163
avoidance, 39, 95, 97, 101, 104, 142, 157
avoidance behavior, 97
awareness, 119

B

background information, 3
bacteria, 121
barometric pressure, 24, 25, 31
barotrauma, 6, 25, 30, 36, 42, 61, 63, 64, 66, 96, 143
barriers, 29
base, 2, 36, 38, 41, 44, 52, 60, 112, 117
behaviors, 7, 27, 87
benefits, 147
bias, 12, 20, 32, 33, 40, 48, 53, 70, 73, 85, 110, 117, 149
biodiversity, 156
biomass, 19, 159
blood, 36
body size, 121
bone(s), 6, 36, 37, 42, 61, 63, 64
bounds, 100, 110, 118
brain, 36
branching, 151
breeding, 8, 46, 49, 62, 83, 101, 112, 116, 117, 132
Bureau of Land Management, 83

C

carbon, 94, 99, 153

carbon dioxide, 94, 99, 153
case study, 95, 129, 150
categorization, 160
category d, 104
CDC, 53
CEC, 77
challenges, 19, 56
citizens, 102
classes, 117
classification, 35
climate, 1, 28, 62, 66, 70, 147
climate change, 1, 66, 147
coal, 66, 94
coefficient of variation, 92, 122
collisions, 6, 8, 9, 13, 30, 32, 33, 39, 41, 46, 49, 56, 59, 60, 63, 66, 72, 77, 78, 85, 93, 94, 95, 96, 98, 102, 104, 108, 109, 111, 117, 118, 132, 144, 147, 155, 166
commercial, 13, 63, 115, 161
communication, 7, 29, 43, 44, 45
communities, 56
community, 8, 16, 37, 59, 62, 80, 97, 163
compensation, 119
complexity, 150
composition, 8, 34, 37, 41, 47, 59, 83
computer, 98
conceptual model, 40
conflict, 73
connectivity, 28, 29, 154
consensus, 88
conservation, vii, 2, 3, 29, 30, 56, 69, 93, 95, 98, 99, 100, 104, 105, 106, 107, 108, 121, 130, 132, 134, 135, 144, 145, 150, 151, 152, 153, 154, 157, 158, 159, 160, 165, 166, 167
construction, 13, 19, 32, 34, 56, 59, 62, 69, 76, 77, 78, 80, 81, 82, 84, 85, 86, 87, 88, 89, 96
consumers, 96
consumption, 1
contamination, 155
control group, 63
cooling, 66
cooperation, 55, 95
coral reefs, 159
correlation(s), 14, 45, 51
cost, 56, 142
Council of Europe, 98, 158
courtship, 28
crop, 9, 41, 42

D

data collection, 148, 149
data set, 13, 31
database, 73, 113, 148, 161
death rate, 69
deaths, 6, 38, 66, 73, 119, 125, 143, 144
decay, 18
decision makers, 93, 102, 107, 120, 130, 139
decomposition, 6, 63, 64
deficiency, 110
degradation, 104, 105
demographic data, 163
demography, 115, 154, 155
Department of Agriculture, 94, 158, 164, 167
Department of Commerce, 152, 164
Department of Energy, 88
Department of the Interior, 88, 92, 94, 164
depth, 138
designers, 100
detectable, 64, 147
detection, 7, 9, 14, 22, 33, 35, 40, 50, 51, 60, 61, 65, 66, 67, 123, 149
deviation, 118
diaphragmatic hernia, 36
displacement, 19, 87, 113, 163
distribution, 58, 60, 70, 73, 85, 103, 104, 111, 112, 114, 116, 117, 118, 121, 126, 129, 144, 162
diversity, 13, 62, 63
DNA, 57
dogs, 12, 18
draft, 66, 165
drawing, 6, 30

E

ecology, 4, 52, 53, 79, 83, 151, 152, 153, 166, 167
ecosystem, 53, 122
edema, 63, 64
education, 19
egg, 136
EIS, 92, 102
electricity, 1, 2, 14, 16, 25, 26, 45, 66, 68, 72, 96, 97, 126, 156
electromagnetic, 74
electromagnetic fields, 74
endangered, 30, 45, 83, 122, 147, 153, 164, 165, 169
energy consumption, 1
environmental conditions, 63
environmental contamination, 155
environmental impact, 54, 92, 102, 155
environmental impact statement, 92, 102
environmental variables, 38
environments, 46, 151, 166
equilibrium, 122
etiology, 63, 161
Europe, 75, 98, 158, 162
evidence, vii, 2, 7, 9, 13, 32, 39, 41, 42, 45, 54, 60, 61, 63, 64, 69, 132, 160
evolution, 151, 166
experimental design, 56
exposure, 106, 111, 139
extinction, 106, 115, 147, 149, 150, 157, 159, 160, 169
extravasation, 63, 64

F

FAA, 17, 20, 41, 42, 48
Fairbanks, 84
farmland, 41
farms, 16, 45, 46, 56, 66, 73, 74, 111, 153, 154, 155, 156, 157, 158, 161, 165
federal government, 45, 120
federal permitting, 19
field trials, 60
fish, 152, 161
Fish and Wildlife Service, 3, 4, 93, 98, 102, 108, 120, 131, 135, 144, 147, 163, 165
fisheries, 159, 161, 162
fishing, 119, 155
flexibility, 95
flight, 39, 42, 57, 72, 98, 145, 146, 157, 167, 168
flights, 73
fluid, 36
food, 27, 28
force, 6, 63, 64
forecasting, 19, 169
fractures, 36, 37, 42, 52, 64
FRE, 5
freezing, 63, 64

G

genus, 8, 37, 38, 62
Geographic Information System (GIS), 41
Georgia, 53
Germany, 73, 74, 98
GIS, 41, 60
google, 4
Google, 4
GPS, 165
grass, 60, 72, 85
grasslands, 23, 158
gravity, 29
greenhouse, 126
grids, 114
growth, 92, 93, 94, 99, 116, 117, 118, 120, 121, 122, 126, 133, 137, 139, 141, 146, 150, 155, 163
growth rate, 92, 93, 116, 117, 118, 120, 122, 133, 137, 139, 141, 146, 155, 163
guidance, 16, 55, 75, 165
guidelines, 19, 35, 56, 57, 77, 102, 120, 129, 130, 157, 160, 165
Gulf Coast, 56

H

Hawaii, 101
health, 53, 122

height, 23, 26, 27, 39, 54, 57, 67, 68, 125, 126, 128
hemorrhage, 36
hemothorax, 36
herpetofauna, 56
high fat, 11, 13, 25
Highlands, 54
history, 16, 28, 109, 111, 113, 121, 124, 133, 147, 157, 160, 163
homes, 45
hot spots, 61
hub, 128
human, 18, 28, 119, 154, 165
hunting, 119, 123
hypothesis, 7, 14, 25, 26, 27, 42, 43, 44, 60, 61

I

identification, 124
IEA, 92, 126
illumination, 24, 25, 31, 67, 68
imagery, 114
images, 89, 90
impact assessment, 102, 155
improvements, vii, 2, 6, 64, 100
incidence, 39, 58
independence, 116
indirect effect, 3, 95, 97, 113, 114
individuals, 10, 38, 39, 52, 55, 61, 94, 101, 109, 116, 118, 120, 133, 150
industry, 27, 55, 69, 89, 95, 139, 143, 151
inferences, 134
information exchange, 19
inguinal, 36
injuries, 6, 25, 30, 35, 36, 47, 52, 69
injury, 6, 36, 61, 63, 64, 96, 162
inner ear, 36, 37
insects, 39, 69, 146
integration, 167
interface, 147
International Energy Agency (IEA), 92, 96, 126, 156
invariants, 160
Iowa, 9, 17, 20, 41, 42, 71, 81, 84
IRI, 92, 104, 107, 108, 113, 114
islands, 13, 50
isotope, 146
issues, vii, 3, 4, 8, 16, 19, 56, 95, 101, 108, 129, 161

J

juveniles, 58, 116, 144

K

kill, 10, 28, 58

L

laceration, 6, 64
landscape, 7, 22, 27, 28, 29, 48, 154
laptop, 38
legislation, 30
legs, 47
lesions, 6, 35, 36, 63, 64
life cycle, 94, 101, 104, 144
light, 88, 114, 146
linear model, 43, 44, 62
longevity, 28
low risk, 120, 138, 143
low temperatures, 13, 50
lying, 39

M

magnitude, 40, 93, 118
majority, 7, 9, 10, 17, 20, 29, 31, 33, 36, 38, 39, 60
mammal(s), 31, 56, 119, 120, 121, 122, 130, 152, 158, 160, 164
management, 19, 51, 56, 61, 100, 106, 115, 119, 122, 123, 124, 131, 142, 148, 153, 158, 160, 162, 164
mapping, 112
marine fish, 162
Marine Mammal Protection Act, 92, 120, 121, 122
Mars, 86

marsh, 34, 35
mass, 146
matrix, 116, 117, 133, 136, 146, 147, 149
median, 62, 127
meta-analysis, 104, 124
metabolism, 28
Mexico, 72, 101
mice, 63, 64
migrants, 48, 57, 72
migration, 7, 8, 12, 13, 14, 17, 22, 23, 25, 27, 28, 29, 31, 32, 43, 45, 46, 49, 50, 51, 57, 60, 63, 70, 71, 72, 73, 84, 111, 112, 146, 161, 167, 169
migration routes, 22
Migratory Bird Treaty Act, 30, 96
Minneapolis, 83
modelling, 159
models, 11, 43, 44, 48, 58, 60, 65, 67, 68, 93, 95, 99, 105, 115, 116, 119, 125, 126, 129, 142, 147, 148, 149, 150, 153, 155, 159, 161, 162, 164, 168, 169
modifications, 123, 129
Montana, 85, 87
mortality rate, 7, 8, 9, 10, 11, 33, 34, 35, 44, 46, 58, 106, 109, 117, 118, 130, 150
museums, 29

N

National Environmental Policy Act(NEPA), 92, 102
National Oceanic and Atmospheric Administration, 152, 164
National Renewable Energy Laboratory(NREL), 55, 85, 86, 88, 92, 125, 127, 156
National Research Council, 85, 97, 160
natural gas, 66
negative effects, 16, 56
NEPA, 92, 102
neurons, 73
New Zealand, 161
NOAA, 152, 164
normal distribution, 118
North America, vii, 2, 3, 4, 16, 17, 19, 27, 29, 45, 53, 54, 61, 68, 69, 70, 92, 110, 112, 114, 132, 147, 151, 155, 156, 157, 158, 160, 161, 162, 163, 165, 166, 168, 169
NREL, 55, 75, 85, 86, 92, 125, 126, 156

O

obstruction, 38
oil, 66, 94, 99, 100, 162
Oklahoma, 9, 10, 17, 60, 72, 85, 143, 169
operations, 3, 15, 25, 26, 40, 65, 81
opportunities, vii, 2, 3, 27, 56, 69, 157, 158
overlap, 144, 146
oversight, 99

P

parallel, 65
parameter estimates, 103, 148
participants, 4
pathway(s), 6, 30, 112, 146
peer review, 5
pelvis, 36
percentile, 121, 122, 126
personal communication, 27
petroleum, 153, 162
pilot study, 12, 18, 73
pitch, 15, 26
plants, 66
pneumothorax, 35, 36
polar, 120
policy, 55, 69, 100, 107, 122, 130
policy issues, 101
policy makers, 55
pollution, 66
ponds, 37, 48, 62
population growth, 92, 93, 99, 116, 117, 118, 120, 121, 122, 133, 137, 139, 141, 146, 155, 163
population size, 3, 29, 92, 93, 98, 102, 103, 104, 106, 109, 110, 116, 118, 119, 120, 121, 122, 123, 131, 133, 138, 139, 141, 144, 149, 150, 158

portfolio, 19
positive correlation, 45
power plants, 66
predator, 58
predictor variables, 44
preparation, 152
prevention, 81
principles, 19, 99
probability, 9, 40, 41, 61, 67, 119, 121, 142
probability distribution, 121
project, 40, 47, 62, 65, 72, 78, 80, 81, 84, 88, 93, 99, 100, 111, 114, 125, 139, 142, 143, 145, 147, 149, 150, 152
protection, 94
public education, 19
public health, 53

R

radar, 57, 74, 84, 146
radio, 63
radius, 44, 61
real time, 39
realism, 115
reality, 119
reasoning, 103
recommendations, 19, 42, 57, 72, 88, 124
recovery, 52, 59, 60, 92, 120, 121, 122, 123, 139, 140, 141, 153
Red List, 108, 144, 156, 157
regression, 31, 32, 48, 128
regression model, 48
regulations, 19, 165
regulatory agencies, 124, 130
reliability, 169
renewable energy, 19, 55, 94, 96, 97, 125, 157
reproduction, 28, 92, 109, 111, 116, 119, 131, 143
reproductive age, 116
requirements, 19
researchers, 55, 98, 100, 109, 111, 125, 149
resource management, 106
resources, 4, 27, 69, 86, 94, 99, 100, 150, 153
response, 16, 28, 68, 94, 103, 105, 106, 114
Richland, 78
risk assessment, 74, 75, 77, 100, 115, 139, 145
risks, 57, 107, 145, 160
routes, 22, 23, 148
rules, 158

S

Sacramento Municipal Utility District, 88
safety, 120
sampling error, 150
scapula, 36
scavengers, 32, 41, 53
science, 72, 100, 112, 142, 158, 160, 164, 165, 169
scientific method, 98
scientific papers, 27
scope, 19, 94
seeding, 18
sensitivity, 40, 98, 111, 154
services, 71
sex, 58, 60, 62, 111, 132
sex ratio, 60, 62, 111, 132
shape, 116, 124
showing, 62, 97, 103, 114, 121, 123, 127, 128, 140, 141, 149
shrubs, 38
signal-to-noise ratio, 150
signs, 6, 35, 42, 52, 63
silver, 7, 8, 11, 13, 14, 17, 20, 22, 23, 24, 25, 33, 35, 41, 43, 44, 45, 46, 47, 49, 50, 51, 52, 53, 54, 55, 59, 62
simulation(s), 40, 98, 119, 121, 122, 124, 126
software, 38, 115
South Dakota, 81
specialization, 159
spin, 15, 21
Spring, 80, 86
stakeholders, 55, 95, 114
standard deviation, 118
state(s), 14, 19, 50, 57, 69, 92, 148, 150, 151, 157, 159, 166

statistics, 8
sternum, 36
stock, 121, 152
storage, 87, 94, 99, 153
stratified sampling, 34
stressors, 95, 147
structure, 8, 32, 37, 42, 44, 80, 100, 101, 150
structuring, 100
styles, 2
subsistence, 123
success rate, 106, 111
surrogates, 57
surveillance, 53
survival, 3, 28, 52, 93, 98, 106, 109, 110, 111, 116, 117, 131, 143, 144, 149, 162
survival rate, 3, 110
sustainability, 159
Sweden, 73
Switzerland, 156
syndrome, 147, 154, 164, 167, 169
synthesis, 3, 4, 5, 16, 18, 19, 30, 154

T

target, 102, 104, 121, 122, 144
target population, 104, 122, 144
teams, 11, 18
techniques, 35, 57, 160
technologies, 146
technology, 100, 146
temperature, 6, 13, 23, 32, 35, 38, 62, 63, 64, 67
temporal variation, 69
Tennessee Valley Authority, 32, 79, 85, 87
testing, 18, 30, 42, 53
threats, 30, 66, 106
time periods, 24, 68, 69, 95
time series, 103, 115, 116, 117, 133, 145, 150
TIR, 7, 38
tissue, 6, 52, 64, 96, 143
total utility, 1
trachea, 63
traits, 65, 111
transmission, 53, 56, 113
transparency, 100
transport, 96
trauma, 6, 36, 42, 52, 63, 64
TRC, 87
treaties, 30
treatment, 15, 26, 52
trial, 11, 18, 63
triggers, 25

U

U.S. Department of Agriculture, 94, 158, 164
U.S. Department of Commerce, 152
U.S. Department of the Interior, 88, 92, 94, 164
U.S. Geological Survey(USGS), 1, 3, 4, 70, 72, 93, 94, 110, 112, 143, 145, 147, 149, 153, 154, 158, 161, 162, 165, 167, 168, 169
United Kingdom, 97, 156
uranium, 66
USGS, 4, 5, 93, 94, 96, 99, 100, 102, 103, 106, 107, 112, 115, 120, 130, 132, 138, 139, 142, 143, 147, 148, 149, 150, 153, 159, 162
UV light, 88

V

variables, 11, 23, 32, 34, 38, 44, 45, 50, 68, 99, 106, 121, 163
variations, 119
vegetation, 7, 18, 32, 34, 41, 44, 45, 56, 58, 60, 157
vegetative cover, 12, 18
vertebrae, 36
vertebrates, 69, 163
victims, 6, 30
vision, 13, 32
visual stimuli, 13, 32
vulnerability, 155

W

walking, 65
Washington, 4, 72, 75, 76, 78, 79, 80, 83, 85, 86, 87, 88, 89, 157, 160, 163
water, 13, 33, 38, 57
weather patterns, 44, 66
wetlands, 48
wildlife, vii, 2, 3, 12, 19, 30, 34, 40, 56, 59, 69, 75, 78, 79, 87, 88, 93, 94, 95, 96, 97, 98, 99, 100, 102, 108, 124, 131, 138, 144, 145, 146, 151, 152, 154, 158, 161, 163
wildlife conservation, 56, 145, 152, 158
wind farm, 16, 45, 46, 47, 56, 58, 61, 63, 65, 66, 71, 72, 73, 74, 79, 84, 85, 111, 113, 153, 154, 155, 156, 157, 158, 162, 165
wind power, 1, 46, 69, 71, 78, 81, 83, 84, 88, 165
wind speed(s), 8, 11, 13, 14, 15, 17, 21, 23, 24, 25, 26, 30, 31, 32, 38, 50, 54, 62, 63, 67, 68
Wisconsin, 2, 6, 9, 34, 35, 77, 80, 81, 86
woodland, 48